CliffsNotes® Biology Quick Review

CliffsNotes® Biology Quick Review

By Kellie Ploeger Cox, Ph.D.

About the Author

Kellie Ploeger Cox, Ph.D., is on the science faculty of Hopkins School located in New Haven, Connecticut.

Editorial

Executive Editor: Greg Tubach
Senior Editor: Christina Stambaugh
Production Editor: Jennifer Freilach
Copy Editor: Lynn Northrup
Technical Editor: Scott Ryan
Proofreader: Donna Wright
Indexer: Potomac Indexing, LLC

CliffsNotes® Biology Quick Review, 4th Edition

Library of Congress Control Number: 2018964153

ISBN: 978-1-957671-23-9

Printed in the United States of America
DOC 10 9 8 7 6 5 4 3 2 1

Published by:
Learneo, Inc.
2000 Seaport Blvd, 3rd Floor
Redwood City, CA 94063
United States of America

Table of Contents

INTRODUCTION

This book provides a concise, comprehensive review of the key concepts of the study of living things. Topics covered in detail include the atoms and molecules that comprise all living organisms; the central dogma of biology; the three central molecules (DNA, RNA, and protein); and the most vital biological processes, such as mitosis, meiosis, and reproduction. Subdisciplines of biology are also thoroughly explained, including taxonomy, anatomy, and ecology. A review of genetics and evolution includes the controversial and groundbreaking theories of Gregor Mendel and Charles Darwin. The biochemical processes of respiration and photosynthesis are discussed in detail for comparison, and several chapters focus on plant structure and function. Due to the increasing importance of recombinant biotechnology and stem cell research, Chapter 11 highlights common molecular biology techniques used to identify genetic disorders and combat disease in humans. Finally, this book provides several chapters reviewing comparative anatomy of humans and other organisms.

CliffsNotes Biology Quick Review, 3rd Edition, includes bare-bones information about a wide variety of biological processes, from microbiology and biochemistry to evolution and ecology.

Why You Need This Book

Can you answer yes to any of these questions?

- Do you need to review the fundamentals of biology fast?
- Do you need a course supplement to aid your study of biology?
- Do you need a concise, comprehensive reference for your biology studies?

If so, then *CliffsNotes Biology Quick Review,* 3rd Edition, is for you!

How to Use This Book

You can use this book in any way that fits your personal style for study and review; you decide what works best for your needs. You can either read the book from cover to cover or just look for the information you want and put the book back on the shelf for later. Here are just a few ways you can search for topics:

- Look for areas of interest in the book's Table of Contents, or use the Index to find specific topics.
- Flip through the book, looking for subject areas at the top of each page.
- Get a glimpse of what you'll gain from a chapter by reading through the Chapter Check-In at the beginning of each chapter.
- Use the Chapter Check-Out at the end of each chapter to gauge your grasp of the important information you need to know.
- Test your knowledge more completely in the Review Questions.
- Use the Glossary to find key terms fast. This book defines new terms and concepts where they first appear in the chapter. If a word is boldfaced, you can find its definition in the Glossary.
- Flip through the book until you find what you're looking for. We organized this book to gradually build on key concepts.

Hundreds of Practice Questions Online!

Go to CliffsNotes.com for hundreds of additional biology practice questions to help you prepare for your next quiz or test. The questions are organized by this book's chapter sections, so it's easy to use the book and then quiz yourself online to make sure you know the subject matter. Visit CliffsNotes.com to test yourself anytime and find other free homework help.

Chapter 1

THE SCIENCE OF BIOLOGY

Chapter Check-In

- ❑ Identifying the characteristics of living things
- ❑ Mapping out the scientific method

Biology is the study of living things. It encompasses the cellular basis of living things, the energy metabolism that underlies the activities of life, and the genetic basis for inheritance in organisms. Biology also includes the study of evolutionary relationships among organisms and the diversity of life on Earth. It considers the biology of microorganisms, plants, and animals, for example, and it brings together the structural and functional relationships that underlie their day-to-day activities. Biology draws on the sciences of chemistry and physics for its foundations and applies the laws of these disciplines to living things.

Many subdisciplines and special areas of biology exist, which can be conveniently divided into practical and theoretical categories. Types of *practical biology* include plant breeding, wildlife management, medical science, and crop production. *Theoretical biology* encompasses such disciplines as *physiology* (the study of the function of living things), *biochemistry* (the study of the chemistry of organisms), *taxonomy* (classification), *ecology* (the study of populations and their interactions with each other and their environments), and *microbiology* (the study of microscopic organisms).

Characteristics of Living Things

Defining a living thing is a difficult proposition, as is defining "life"—that property possessed by living things. However, a living thing possesses certain properties that help define what life is.

Cells

The **cell** is the smallest unit of life. A living thing is composed of one or more cells. These units, generally too small to be seen with the naked eye, are organized into tissues. A *tissue* is a series of cells that accomplish a shared function. Tissues, in turn, form *organs,* such as the stomach and kidneys. A number of organs working together compose an *organ system.* An *organism* is a complex series of various organ systems.

Metabolism

Living things exhibit a rapid turnover of chemical materials, which is referred to as **metabolism.** Metabolism involves exchanges of chemical matter with the external environment and extensive transformations of organic matter within the cells of a living organism. Metabolism generally involves the release or use of chemical energy. Nonliving things do not display metabolism.

Responsiveness

All living things are able to respond to stimuli in the external environment. Living things respond to changes in light, heat, sound, and chemical and mechanical contact. For example, to detect stimuli, animals have means for receiving information, such as eyes, ears, and taste buds. Plants can detect changes in day length and adjust their growth accordingly.

To respond effectively to changes in the environment, an organism must coordinate its responses. A system of nerves and a number of chemical regulators called **hormones** coordinate activities within animals. The animal responds to the stimuli by means of a number of effectors, such as muscles and glands. Energy is generally used in the process.

Organisms change their behavior in response to changes in the surrounding environment. For example, an organism may move in response to its environment. Responses such as this occur in definite patterns and make up the behavior of an organism. The behavior is active, not passive; an animal responding to a stimulus is different from a stone rolling down a hill. Living things display **responsiveness;** nonliving things do not.

Growth

Growth requires an organism to take in material from the environment and organize the material into its own structures. To accomplish growth, an organism expends some of the energy it acquires during metabolism. An organism has a pattern for accomplishing the building of growth structures.

During growth, a living organism transforms material that is unlike itself into materials that are like it. A person, for example, digests a meal of meat and vegetables and transforms the chemical material into macromolecules needed for constructing more cells and tissues. A nonliving organism does not display this characteristic.

Reproduction

A living thing has the ability to produce copies of itself by the process known as *reproduction.* These copies are made while the organism is still living. For some simple organisms, reproduction involves only one parent, and the offspring are genetically identical to the parent. This is called *asexual reproduction.* For example, a bacterium reproduces by splitting in half, a process called *binary fission.* Some plants can create clones of themselves if a part of the parent plant breaks off and is able to form roots and grow independently. More complex organisms engage in a type of reproduction called *sexual reproduction,* in which two parents contribute to the formation of a new individual. During this process, a new combination of traits can be produced. Sexual reproduction is a major source of genetic variation in populations of organisms.

Evolution

Living organisms have the ability to adapt to their environment through the process of **evolution.** During evolution, changes occur in populations, and the organisms in the population become better able to metabolize, respond, and reproduce. They acquire traits better suited to their environment that their ancestors did not have. This is how a species evolves.

Evolution also results in a greater variety of species than existed in previous eras. This proliferation of populations of organisms is unique to living things.

Ecology

The environment influences the living things that it surrounds. *Ecology* is the study of relationships between organisms and their relationships with their environment. Both *biotic factors* (living things) and *abiotic factors* (nonliving things) can alter the environment. Rain and sunlight are nonliving components, for example, that greatly influence the environment. Living things may migrate or hibernate if the environment becomes difficult to live in.

Scientific Method

Biology is one of the major sciences. Scientists have acquired biological knowledge through a process known as the **scientific method.** The steps of the scientific method make up an orderly way of gaining information about the biological world. Sometimes the knowledge gained is useful in solving particular problems; other times it is simply of interest without any practical application at the time.

The scientific method requires a systematic search for information by observation and experimentation. The basic steps of the scientific method are stating a problem based on observations, developing a research question or questions, forming a hypothesis, experimenting to test the hypothesis, collecting information, recording and analyzing data, and forming a conclusion.

Observation

The first step in the scientific method is stating a problem based on observation. In this stage, the scientist recognizes that something has happened and that it occurs repeatedly. Therefore, the scientist formulates a question or states a problem for investigation. The next step in the scientific method is to explore resources that may have information about that question or problem. Here, the scientist conducts a literature review and interacts with other scientists to develop knowledge about the question at hand.

Hypothesis, experimentation, and analysis

Next, a **hypothesis** is formed, meaning that the scientist proposes a possible solution to the question, realizing that the answer could be incorrect. The scientist tests the hypothesis through experiments that include experimental and control groups. Data from the experiments are collected, recorded, and analyzed.

Conclusion

After analyzing the data, the scientist draws a conclusion. A valid conclusion must be based on the facts observed in the experiments. If the data from repeated experiments support the hypothesis, the scientist publishes the hypothesis and experimental data for other scientists to review and discuss.

Theory and scientific law

Other scientists not only may repeat the experiments, but also may carry out additional experiments to challenge the findings. If the hypothesis is tested and confirmed often enough, the scientific community calls the hypothesis a **theory.** Then numerous additional experiments test the theory using rigorous experimental methods. Repeated challenges to the theory are presented. If the results continue to support the theory and are found to be universally applicable, the theory is elevated to a scientific law. A *scientific law* is a universal, uniform, or constant fact of nature. Sir Isaac Newton's law of gravity is an example of a scientific law.

Chapter Check-Out

Questions

1. List the steps of the scientific method.
2. Describe the difference between a theory and a scientific law.

Answers

1. Making observations, developing a question, forming a hypothesis, experimenting to test the hypothesis, collecting information, recording and analyzing data, and forming a conclusion.
2. A hypothesis that is confirmed often enough is called a theory; a theory can still be challenged and tested. Once a theory is elevated to a scientific law, it is considered a universal truth that no longer requires verification.

Chapter 2

THE CHEMICAL BASIS OF LIFE

Chapter Check-In

- ❑ Understanding basic chemical principles
- ❑ Comparing organic compounds

For many centuries, biology was the study of the natural world. Biologists searched for unidentified plants and animals, classified them, and studied their anatomy and how they acted in nature. Then, in the 1700s, scientists discovered the chemical and physical bases of living things. They soon realized that the chemical organization of all living things is remarkably similar.

Elements

All living things on Earth are composed of fundamental building blocks of matter called **elements.** More than 100 elements are known to exist, including those that are man-made. An element is a substance that cannot be chemically decomposed. Oxygen, iron, calcium, sodium, hydrogen, carbon, and nitrogen are examples of elements.

Atoms

Each element is composed of one particular kind of atom. An **atom** is the smallest part of an element that can enter into combinations with atoms of other elements.

Atoms consist of positively charged particles called *protons* surrounded by negatively charged particles called *electrons.* A third type of particle, a *neutron,* has no electrical charge; it has the same weight as a proton. Protons and neutrons adhere tightly to form the dense, positively charged nucleus of the atom. Electrons spin around the nucleus.

The electron arrangement in an atom plays an essential role in the chemistry of the atom. Atoms are most stable when their outer shell of electrons has a full quota. The first electron shell has a maximum of two electrons. The second and all other outer shells have a maximum of eight electrons. Atoms tend to gain or lose electrons until their outer shells have a stable arrangement. The gaining or losing of electrons, or the sharing of electrons, contributes to the chemical reactions in which an atom participates.

Molecules

Most of the compounds of interest to biologists are composed of units called molecules. A **molecule** is a precise arrangement of atoms held together by chemical bonds, and a **compound** is a molecule that contains atoms of more than one element. A molecule may be composed of two or more atoms of the same element, as in oxygen gas (O_2), or it may be composed of atoms from different elements. The arrangements of the atoms in a molecule account for the properties of a compound. The molecular weight is equal to the atomic weights of the atoms in the molecule.

The atoms in molecules may be joined to one another by various linkages called *bonds.* One example of a bond is an *ionic bond,* which is formed when the electrons of one atom transfer to a second atom. This creates electrically charged atoms called *ions.* The electrical charges cause the ions to be attracted to one another, and the attraction forms the ionic bond.

A second type of linkage is a *covalent bond.* A covalent bond forms when two atoms share one or more electrons with one another. For example, as shown in Figure 2-1, oxygen shares its electrons with two hydrogen atoms, and the resulting molecule is water (H_2O). Nitrogen shares its electrons with three hydrogen atoms, and the resulting molecule is ammonia (NH_3). If one pair of electrons is shared, the bond is a single bond; if two pairs are shared, it is a double bond.

Acids and Bases

Acids are chemical compounds that release hydrogen ions (H+) when placed in water. For example, when hydrogen chloride is placed in water, it releases its hydrogen ions and the solution becomes hydrochloric acid.

Figure 2-1 Formation of a covalent bond in water and ammonia molecules. In each molecule, the second shell fills with eight electrons.

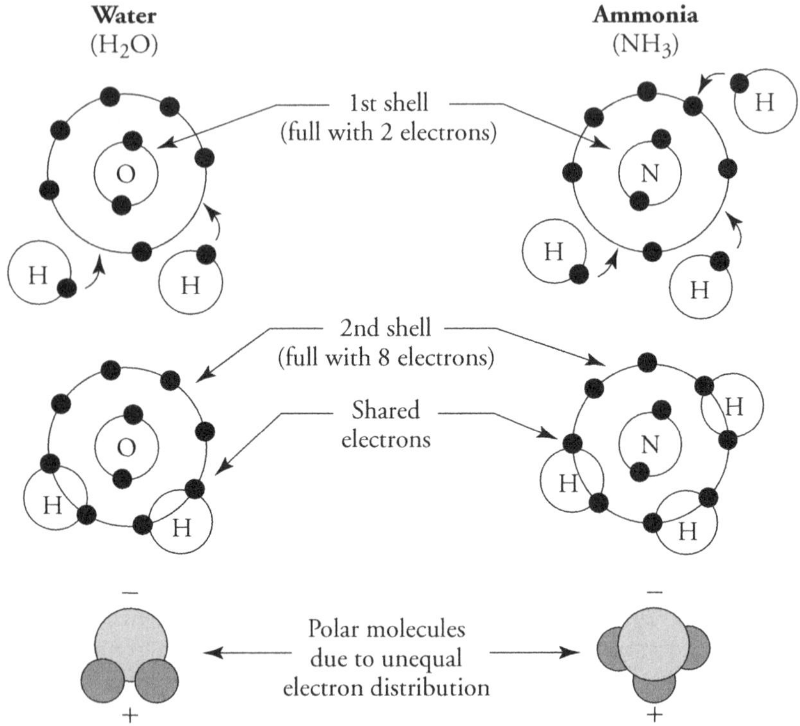

Bases are chemical compounds that attract hydrogen atoms when they are placed in water. An example of a base is sodium hydroxide (NaOH). When this substance is placed in water, it attracts hydrogen ions, and a basic (or alkaline) solution results as hydroxyl (–OH) ions accumulate.

Organic Compounds

The chemical compounds of living things are known as *organic compounds* because of their association with organisms and because they are carbon-containing compounds. Organic compounds, which are the compounds associated with life processes, are the subject matter of organic chemistry. Among the numerous types of organic compounds, four major categories are found in all living things: carbohydrates, lipids, proteins, and nucleic acids.

Carbohydrates

Almost all organisms use **carbohydrates** as sources of energy. In addition, some carbohydrates serve as structural materials. Carbohydrates are molecules composed of carbon, hydrogen, and oxygen; the ratio of hydrogen atoms to oxygen and carbon atoms is 2:1.

Simple carbohydrates, commonly referred to as *sugars,* can be **monosaccharides** if they are composed of single molecules, or **disaccharides** if they are composed of two molecules. The most important monosaccharide is **glucose,** a carbohydrate with the molecular formula $C_6H_{12}O_6$. Glucose is the basic form of fuel in living things. In multicellular organisms, it is soluble and is transported by body fluids to all cells, where it is metabolized to release its energy. Glucose is the starting material for cellular respiration, and it is the main product of photosynthesis (see Chapters 5 and 6).

Three important disaccharides are also found in living things: maltose, sucrose, and lactose. Maltose is a combination of two glucose units covalently linked. The table sugar sucrose is formed by linking glucose to another monosaccharide called *fructose.* Figure 2-2 shows that in the

Figure 2-2 Glucose and fructose molecules combine to form the disaccharide sucrose.

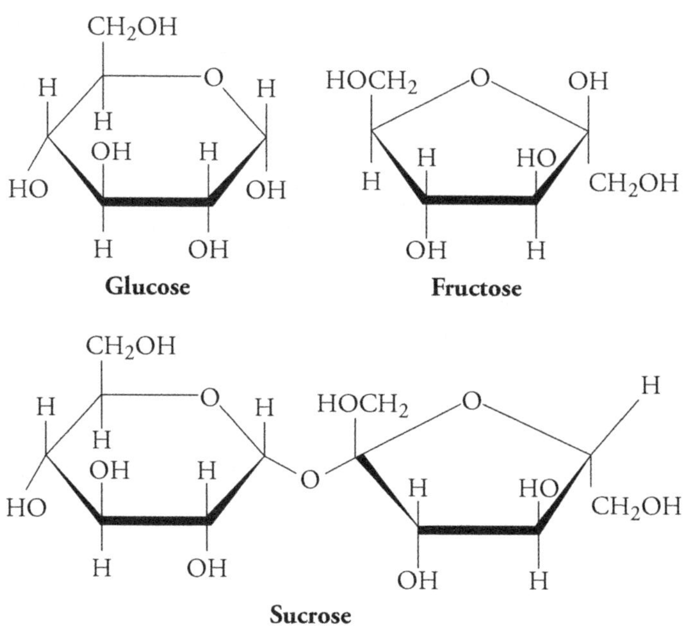

synthesis of sucrose, a water molecule is produced. The process is therefore called a *dehydration reaction.* The reversal of the process is *hydrolysis,* a process in which the molecule is split and water is added. Lactose is composed of glucose and galactose units.

Complex carbohydrates are known as **polysaccharides.** Polysaccharides are formed by linking innumerable monosaccharides. Among the most important polysaccharides is starch, which is composed of hundreds or thousands of glucose units linked to one another. Starch serves as the energy storage form for plants. Much of the world's human population satisfies its energy needs with starch from rice, wheat, corn, and potatoes.

Two other important polysaccharides are glycogen and cellulose. **Glycogen** is also composed of thousands of glucose units, but the units are bonded in a different pattern than in starch. Glycogen is the form in which glucose is stored in the human liver. **Cellulose** is used primarily as a structural carbohydrate in plant cell walls. It is also composed of glucose units, but the units cannot be released from one another except by a few species of organisms (humans do not have the enzymes necessary to digest cellulose). Wood is composed chiefly of cellulose, and cotton fabric and paper are commercial cellulose products.

Lipids

Lipids are organic molecules composed of carbon, hydrogen, and oxygen atoms. The ratio of hydrogen atoms to oxygen atoms is much higher in lipids than in carbohydrates. Lipids include steroids (the material of which many hormones are composed), waxes, and **fats.**

Fat molecules are composed of a glycerol molecule and three molecules of fatty acids (see Figure 2-3). A glycerol molecule contains three hydroxyl (–OH) groups. A **fatty acid** is a long chain of carbon atoms (from 4 to 24) with a carboxyl (–COOH) group at one end. The fatty acids in a fat may all be alike, or they may all be different. They are bound to the glycerol molecule by a process that involves the removal of water (dehydration reaction).

Certain fatty acid molecules have one or more double bonds between carbons. Fats that include these molecules are *unsaturated fats.* Other fatty acids have no double bonds, and only single bonds connect carbons. Fats that include these fatty acids are *saturated fats.* In most human health situations, the consumption of unsaturated fats is preferred to the consumption of saturated fats.

Figure 2-3 A fat molecule is constructed by combining a glycerol molecule with three fatty acid molecules. (Two saturated fatty acids and one unsaturated fatty acid are shown for comparison.) The constructed molecule is at the bottom.

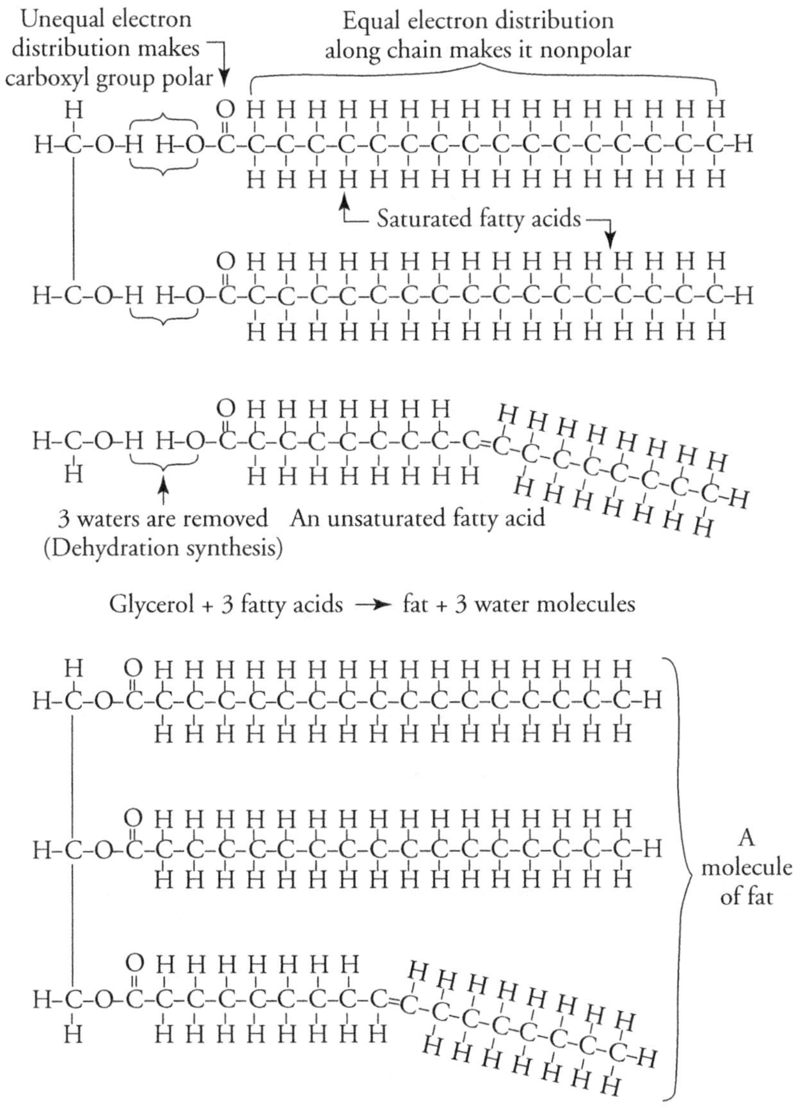

Fats stored in cells usually form clear oil droplets called *globules* because fats do not dissolve in water. Plants often store fats in their seeds, and animals store fats in large, clear globules in the cells of adipose tissue. The fats in adipose tissue contain much concentrated energy. Hence, they serve as a reserve energy supply to the organism. The enzyme lipase breaks down fats into fatty acids and glycerol in the human digestive system.

Proteins

Proteins, among the most complex of all organic compounds, are composed of **amino acids** (see Figure 2-4), which contain carbon, hydrogen, oxygen, and nitrogen atoms. Certain amino acids also have sulfur atoms, phosphorus, or other trace elements such as iron or copper.

Many proteins are immense and extremely complex. However, all proteins are composed of long chains of relatively simple amino acids. There are 20 kinds of amino acids. Each amino acid (see the left illustration in Figure 2-4) has an amino ($-NH_2$) group, a carboxyl ($-COOH$) group, and a group of atoms called an –R group (where R stands for *radical*). The amino acids differ depending on the nature of the –R group, as shown in the middle illustration of Figure 2-4. The –R groups are categorized based on their chemical properties (e.g., acidic, charged, nonpolar, and polar). The property of the side chain has an impact on the three-dimensional folding and function of the overall protein. Examples of amino acids are alanine, valine, glutamic acid, tryptophan, tyrosine, and histidine.

Figure 2-4 The structure and chemistry of amino acids. When two amino acids are joined in a dipeptide, the –OH of one amino acid is removed, and the –H of the second is removed. So, water is removed. A dipeptide bond (right) forms to join the amino acids together.

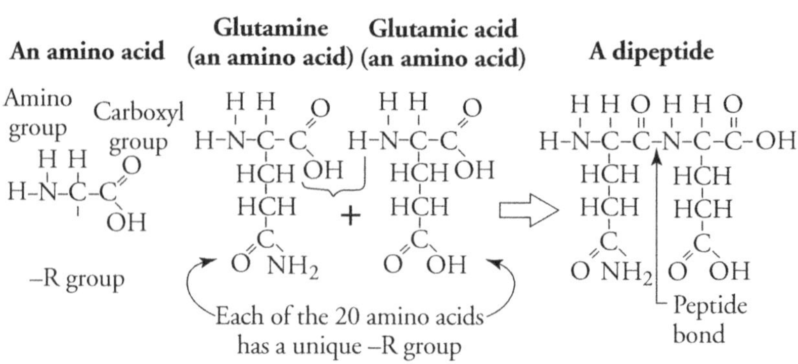

The removal of water molecules links amino acids to form a protein. The process is called *dehydration synthesis,* and a by-product of the synthesis is water. The links forged between the amino acids are *peptide bonds,* and small proteins are often called **peptides.**

All living things depend on proteins for their existence. Proteins are the major molecules from which living things are constructed. Certain proteins are dissolved or suspended in the watery substance of the cells, while others are incorporated into various structures of the cells. Proteins are also found as supporting and strengthening materials in tissues outside of cells. Bone, cartilage, tendons, and ligaments are all composed of proteins.

One essential function of proteins is as an enzyme. **Enzymes** catalyze the chemical reactions that take place within cells. They are not used up in a reaction; rather, they remain available to catalyze succeeding reactions.

Every species manufactures proteins unique to that species. The information for synthesizing the unique proteins is located in the nucleus of the cell. The **genetic code** specifies the amino acid sequence in proteins. Hence, the genetic code regulates the chemistry taking place within a cell. Proteins also can serve as a reserve source of energy for the cell. When the amino group is removed from an amino acid, the resulting compound is energy-rich.

Nucleic acids

Like proteins, **nucleic acids** are very large molecules. The nucleic acids are composed of smaller units called **nucleotides.** Each nucleotide contains a carbohydrate molecule (sugar), a phosphate group, and a nitrogen-containing molecule that, because of its properties, is a **nitrogenous base.**

Living organisms have two important nucleic acids. One type is **deoxyribonucleic acid,** or **DNA.** The other is **ribonucleic acid,** or **RNA.** DNA is found primarily in the nucleus of the cell, while RNA is found in both the nucleus and the *cytoplasm,* a semiliquid substance that composes the volume of the cell (see Chapter 3).

DNA and RNA differ from one another in their components. DNA contains the carbohydrate deoxyribose, while RNA has ribose. In addition, DNA contains the base thymine, while RNA has uracil. The structure of DNA and its importance in cell life is explored in Chapter 10.

Chapter Check-Out

Questions

1. Which of the following is NOT a component of an atom?
   a. neutron
   b. proton
   c. electron
   d. element
2. Which of the following is NOT an example of a carbohydrate?
   a. maltose
   b. glycogen
   c. lipid
   d. glucose
3. The building blocks of proteins are ______.
   a. DNA
   b. amino acids
   c. enzymes
   d. peptides
4. True or False: The structure of DNA contains the sugar *ribose.*

Answers

1. d
2. c
3. b
4. False

Chapter 3
THE BIOLOGY OF CELLS

Chapter Check-In

- ❑ Differentiating prokaryotic and eukaryotic cells
- ❑ Recognizing the organelles of living cells
- ❑ Moving materials through the plasma membrane

One of the basic tenets of biology is that all living things are composed of one or more cells. Some organisms consist of a single cell, while others have multiple cells organized into tissues, and tissues organized into organs. In many living things, organs function together as an organ system. However, even in these complex organisms, the basic biology revolves around the activities of the cell.

One of the first scientists to observe cells was Englishman Robert Hooke. In the mid-1600s, Hooke examined a thin slice of cork through the newly developed microscope. The microscopic compartments in the cork impressed him and reminded him of rooms in a monastery, known as cells. He, therefore, referred to the units as **cells.** Later in that century, Anton van Leeuwenhoek, a Dutch merchant, made further observations of plant, animal, and microorganism cells. In 1838, German botanist Matthias Schleiden proposed that all plants are composed of cells. A year later, his colleague, anatomist Theodor Schwann, concluded that all animals are also composed of cells. In 1858, biologist Rudolf Virchow proposed that all living things are made of cells and that all cells arise from preexisting cells. These premises have come down to us as the *cell theory.*

The Structure of Prokaryote and Eukaryote Cells

During the 1950s, scientists developed the concept that all organisms may be classified as **prokaryotes** or **eukaryotes.** The cells of all prokaryotes and eukaryotes possess two basic features: a plasma membrane, also called a cell membrane, and cytoplasm. However, the cells of prokaryotes are simpler than those of eukaryotes. For example, prokaryotic cells lack a nucleus, while eukaryotic cells have a nucleus. Prokaryotic cells lack internal cellular bodies (organelles), while eukaryotic cells possess them. Examples of prokaryotes are bacteria and **archaea.** Examples of eukaryotes are protists, fungi, plants, and animals (everything except prokaryotes).

Plasma membrane

All prokaryote and eukaryote cells have plasma membranes. The **plasma membrane** (also known as the *cell membrane*) is the outermost cell surface, which separates the cell from the external environment. The plasma membrane is composed primarily of proteins and lipids, specifically phospholipids. The lipids occur in two layers (a *bilayer*). Proteins embedded in the bilayer appear to float within the lipid, so the membrane is constantly in flux. The membrane is therefore referred to as a *fluid mosaic structure.* Within the fluid mosaic structure, proteins carry out most of the membrane's functions.

The "Movement through the Plasma Membrane" section later in this chapter describes the process by which materials pass between the interior and exterior of a cell.

Cytoplasm and organelles

All prokaryote and eukaryote cells also have **cytoplasm** (or **cytosol**), a semiliquid substance that composes the volume of a cell. Essentially, cytoplasm is the gel-like material enclosed by the plasma membrane.

Within the cytoplasm of eukaryote cells are a number of membrane-bound bodies called **organelles** ("little organs") that provide a specialized function within the cell.

One example of an organelle is the **endoplasmic reticulum (ER).** The ER is a series of membranes extending throughout the cytoplasm of eukaryotic cells. In some places, the ER is studded with submicroscopic bodies called **ribosomes.** This type of ER is called **rough ER.** In other places, there are no ribosomes. This type of ER is called **smooth ER.** The rough ER is the site of protein synthesis in a cell because it contains ribosomes; however, the smooth ER lacks ribosomes and is responsible for producing lipids. Within the ribosomes, amino acids are actually bound together to form proteins. *Cisternae* are spaces within the folds of the ER membranes.

Another organelle is the **Golgi apparatus** (also called *Golgi body*). The Golgi apparatus is a series of flattened sacs, usually curled at the edges. In the Golgi apparatus, the cell's proteins and lipids are processed and packaged before being sent to their final destination. To accomplish this function, the outermost sac of the Golgi apparatus often bulges and breaks away to form droplike vesicles known as *secretory vesicles.*

An organelle called the *lysosome* (see Figure 3-1) is derived from the Golgi apparatus. It is a droplike sac of **enzymes** in the cytoplasm. These enzymes are used for digestion within the cell. They break down particles

Figure 3-1 The components of an idealized eukaryotic cell. The diagram shows the relative sizes and locations of the cell parts.

of food taken into the cell and make the products available for use; they also help break down old cell organelles. Enzymes are also contained in a cytoplasmic body called the **peroxisome.**

The organelle that releases quantities of energy to form adenosine triphosphate (ATP) is the **mitochondrion** (the plural form is *mitochondria*). Because mitochondria are involved in energy release and storage, they are called the "powerhouses of the cells."

Green plant cells, for example, contain organelles known as **chloroplasts,** which function in the process of photosynthesis. Within chloroplasts, energy from the sun is absorbed and transformed into the energy of carbohydrate molecules. Plant cells specialized for photosynthesis contain large numbers of chloroplasts, which are green because the chlorophyll pigments within the chloroplasts are green. Leaves of a plant contain numerous chloroplasts. Plant cells not specializing in photosynthesis (for example, root cells) are not green.

An organelle found in mature plant cells is a large, fluid-filled central **vacuole.** The vacuole may occupy more than 75 percent of the plant cell. In the vacuole, the plant stores nutrients, as well as toxic wastes. Pressure within the growing vacuole may cause the cell to swell.

The **cytoskeleton** is an interconnected system of fibers, threads, and interwoven molecules that give structure to the cell. The main components of the cytoskeleton are microtubules, microfilaments, and intermediate filaments. All are assembled from subunits of protein.

The **centriole** organelle is a cylinder-like structure that occurs in pairs. Centrioles function in cell division.

Many cells have specialized cytoskeletal structures called flagella and cilia. *Flagella* are long, hairlike organelles that extend from the cell, permitting it to move. In prokaryotic cells, such as bacteria, the flagella rotate like the propeller of a motorboat. In eukaryotic cells, such as certain protozoa and sperm cells, the flagella whip about and propel the cell. *Cilia* are shorter and more numerous than flagella. In moving cells, the cilia wave in unison and move the cell forward. Paramecium is a well-known ciliated protozoan. Cilia are also found on the surface of several types of cells, such as those that line the human respiratory tract.

Nucleus

Prokaryotic cells lack a **nucleus;** the word *prokaryotic* means "primitive nucleus." Eukaryotic cells, on the other hand, have a distinct nucleus.

The nucleus of eukaryotic cells is composed primarily of protein and **deoxyribonucleic acid,** or **DNA.** The DNA is tightly wound around special proteins called **histones;** the mixture of DNA and histone proteins is called **chromatin.** The chromatin is folded even further into distinct threads called **chromosomes.** Functional segments of the chromosomes are referred to as **genes.** Approximately 21,000 genes are located in the nucleus of all human cells.

The *nuclear envelope,* an outer membrane, surrounds the nucleus of a eukaryotic cell. The nuclear envelope is a double membrane, consisting of two lipid layers (similar to the plasma membrane). Pores in the nuclear envelope allow the internal nuclear environment to communicate with the external nuclear environment.

Within the nucleus are two or more dense organelles referred to as **nucleoli** (the singular form is *nucleolus*). In nucleoli, submicroscopic particles known as *ribosomes* are assembled before their passage out of the nucleus into the cytoplasm.

Although prokaryotic cells have no nucleus, they do have DNA. The DNA exists freely in the cytoplasm as a closed loop. It has no protein to support it and no membrane covering it. A bacterium typically has a single looped chromosome.

Cell wall

Many kinds of prokaryotes and eukaryotes contain a structure outside the cell membrane called the **cell wall.** With only a few exceptions, all prokaryotes have thick, rigid cell walls that give them their shape. Among the eukaryotes, some protists, and all fungi and plants, have cell walls. Cell walls are not identical in these organisms, however. In fungi, the cell wall contains a polysaccharide called *chitin.* Plant cells, in contrast, have no chitin; their cell walls are composed exclusively of the polysaccharide cellulose.

Cell walls provide support and help cells resist mechanical pressures, but they are not solid, so materials are able to pass through rather easily. Cell walls are not selective devices, as plasma membranes are.

Movement through the Plasma Membrane

In order for the cell cytoplasm to communicate with the external environment, materials must be able to move through the plasma membrane. This movement occurs through several mechanisms.

Diffusion

One method of movement through the membrane is **diffusion.** Diffusion is the movement of molecules from a region of higher concentration to one of lower concentration and does not require energy. This movement occurs because the molecules are constantly colliding with one another. The net movement of the molecules is away from the region of high concentration to the region of low concentration. This is referred to as moving down their **concentration gradient.** A drop of dye placed in a beaker of water illustrates diffusion as the dye molecules spread out and color the water.

Osmosis

Another method of movement across the membrane is osmosis. **Osmosis** is the diffusion of water from a region of higher concentration (of water) to one of lower concentration (of water). Osmosis occurs across a membrane that is semipermeable. A semipermeable membrane lets only certain molecules pass through while keeping other molecules out. Osmosis is a type of diffusion, so it does not require energy.

Facilitated diffusion

A third mechanism for movement across the plasma membrane is **facilitated diffusion.** Certain proteins in the membrane assist facilitated diffusion by permitting only certain molecules to pass across the membrane. The proteins encourage movement in the direction that diffusion would normally take place, from a region with a higher concentration of molecules to a region of lower concentration.

Active transport

A fourth method for movement across the membrane is **active transport.** When active transport is taking place, a protein channel moves a certain material across the membrane from a region of lower concentration to a region of higher concentration. Because this movement is happening against the concentration gradient, the cell must expend energy that is usually derived from a substance called adenosine triphosphate, or ATP (see Chapter 4). An example of active transport occurs in human nerve cells: Sodium ions are constantly transported out of the cell into the external fluid bathing the cell, a region of high concentration of sodium. (This transport of sodium sets up the nerve cell for the impulse that will occur within it later.)

Endocytosis and exocytosis

The final mechanism for movement across the plasma membrane into the cell is **endocytosis,** a process in which a small patch of plasma membrane encloses particles or tiny volumes of fluid that are at or near the cell surface. The membrane enclosure then sinks into the cytoplasm and pinches off from the membrane, forming a vesicle that moves into the cytoplasm. When the vesicle contains solid particulate matter, the process is called **phagocytosis.** When the vesicle contains droplets of fluid, the process is called **pinocytosis.** Along with the other mechanisms for transport across the plasma membrane, endocytosis ensures that the internal cellular environment will be able to exchange materials with the external environment and that the cell will continue to thrive and function. **Exocytosis** is the reverse of endocytosis; internally produced substances are enclosed in vesicles and fuse with the cell membrane, releasing the contents to the exterior of the cell. Both endocytosis and exocytosis require energy to occur.

Chapter Check-Out

Questions

1. The nucleus contains which of the following?

- **a.** cytoplasm
- **b.** DNA
- **c.** endoplasmic reticulum
- **d.** Golgi apparatus

2. Energy, or ATP, is produced in what organelle?

- **a.** nucleus
- **b.** ribosome
- **c.** mitochondria
- **d.** vacuole

3. The cytoskeleton is composed of all of the following structures EXCEPT ______.

- **a.** intermediate filaments
- **b.** microtubules
- **c.** cilia
- **d.** microfilaments

4. Movement across the plasma membrane from lower to higher concentration requiring energy is an example of which of the following processes?
   a. diffusion
   b. osmosis
   c. facilitated diffusion
   d. active transport

Answers

1. b
2. c
3. c
4. d

Chapter 4
CELLS AND ENERGY

Chapter Check-In

- ❑ Observing the laws of thermodynamics
- ❑ Catalyzing chemical reactions
- ❑ Defining the energy unit of a cell

Life can exist only where molecules and cells remain organized. All cells need energy to maintain organization. Physicists define energy as the ability to do work; in this case, the work is the continuation of life itself.

Energy has been expressed in terms of reliable observations known as the *laws of thermodynamics.* There are two such laws. The first law of thermodynamics states that energy can neither be created nor destroyed. This law implies that the total amount of energy in a closed system (for example, the universe) remains constant. Energy neither enters nor leaves a closed system.

Within a closed system, energy can change, however. For example, the chemical energy in gasoline is released when the fuel combines with oxygen and a spark ignites the mixture within a car's engine. The gasoline's chemical energy is changed into heat energy, sound energy, and the energy of motion.

The second law of thermodynamics states that the amount of *available* energy in a closed system is decreasing constantly. Energy becomes unavailable for use by living things because of **entropy,** which is the degree of disorder or randomness of a system. The entropy of any closed system is constantly increasing. In essence, any closed system tends toward disorganization.

Unfortunately, the transfers of energy in living systems are never completely efficient. Every body movement, every thought, and every chemical reaction in the cells involves a shift of energy and a measurable decrease of energy available to do work in the process. For this reason, considerably more energy must be taken into the system than is necessary to carry out the actions of life.

Chemical Reactions

Most chemical compounds do not combine with one another automatically, nor do chemical compounds break apart automatically. The great majority of the chemical reactions that occur within living things must be energized. This means that the atoms of a molecule must be separated by energy put into the system. The energy forces apart the atoms in the molecules and allows the reaction to take place.

To initiate a chemical reaction, a type of "spark," referred to as the *energy of activation,* is needed. For example, hydrogen and oxygen can combine to form water at room temperature, but the reaction requires activation energy.

Any chemical reaction in which energy is released is called an **exergonic reaction.** In an exergonic chemical reaction, the products end up with less energy than the reactants. Other chemical reactions are **endergonic reactions.** In endergonic reactions, energy is obtained and trapped from the environment. The products of endergonic reactions have more energy than the reactants taking part in the chemical reaction. For example, plants carry out the process of photosynthesis, in which they trap energy from the sun to form carbohydrates (see Chapter 5).

The activation energy needed to spark an exergonic or endergonic reaction can be heat energy or chemical energy. Reactions that require activation energy can also proceed in the presence of biological catalysts. *Catalysts* are substances that speed up chemical reactions but remain unchanged themselves. Catalysts work by lowering the required amount of activation energy for the chemical reaction. For example, hydrogen and oxygen combine with one another in the presence of platinum. In this case, platinum is the catalyst. In biological systems, the most common catalysts are protein molecules called enzymes. Enzymes are absolutely essential if chemical reactions are to occur in cells.

Enzymes

The chemical reactions in all cells of living things operate in the presence of biological catalysts called **enzymes.** Because a particular enzyme catalyzes only one reaction, there are thousands of different enzymes in a cell catalyzing thousands of different chemical reactions. The substance changed or acted on by an enzyme is its **substrate.** The products of a chemical reaction catalyzed by an enzyme are *end products.*

All enzymes are composed of proteins. (Proteins are chains of amino acids; see Chapter 2.) When an enzyme functions, a key portion of the enzyme, called the **active site,** interacts with the substrate. The active site closely matches the molecular configuration of the substrate. After this interaction has taken place, a change in shape in the active site places a physical stress on the substrate. This physical stress aids the alteration of the substrate and produces the end products. During the time the active site is associated with the substrate, the combination is referred to as the *enzyme-substrate complex.* After the enzyme has performed its work, the product or products are released from the enzyme's active site. The enzyme is then free to function in another chemical reaction.

Enzyme-catalyzed reactions occur extremely fast. They happen about a million times faster than uncatalyzed reactions. With some exceptions, the names of enzymes end in "–ase." For example, the enzyme that breaks down hydrogen peroxide to water and hydrogen is catalase. Other enzymes include amylase, hydrolase, peptidase, and kinase.

The rate of an enzyme-catalyzed reaction depends on a number of factors, such as the concentration of the substrate, the acidity and temperature of the environment, and the presence of other chemicals. At higher temperatures, enzyme reactions occur more rapidly, but only up to a point. Because enzymes are proteins, excessive amounts of heat can change their structures, rendering them inactive. An enzyme altered by heat is said to be *denatured.*

Enzymes work together in metabolic pathways. A *metabolic pathway* is a sequence of chemical reactions occurring in a cell. A single enzyme-catalyzed reaction may be one of multiple reactions in a metabolic pathway. Metabolic pathways may be of two general types: catabolic and anabolic. Catabolic pathways involve the breakdown or digestion of large, complex molecules. The general term for this process is **catabolism.** Anabolic pathways involve the synthesis of large molecules, generally by joining smaller molecules together. The general term for this process is **anabolism.**

Many enzymes are assisted by chemical substances called **cofactors.** Cofactors may be ions or molecules associated with an enzyme and are required in order for a chemical reaction to take place. Ions that might operate as cofactors include those of iron, manganese, and zinc. Organic molecules acting as cofactors are referred to as **coenzymes.** Examples of coenzymes are NAD and FAD (see the "ATP Production" section below).

Adenosine Triphosphate (ATP)

The chemical substance that serves as the currency of energy in a cell is **adenosine triphosphate (ATP).** ATP is referred to as currency because it can be "spent" in order to make chemical reactions occur. The more energy required for a chemical reaction, the more ATP molecules must be spent.

Virtually all forms of life use ATP, a nearly universal molecule of energy transfer. The energy released during catabolic reactions is stored in ATP molecules. In addition, the energy trapped in anabolic reactions (such as photosynthesis) is trapped in ATP molecules.

An ATP molecule consists of three parts. One part is a double ring of carbon and nitrogen atoms called *adenine.* Attached to the adenine molecule is a small five-carbon carbohydrate called *ribose.* Attached to the ribose molecule are three phosphate units linked together by covalent bonds.

The covalent bonds that unite the phosphate units in ATP are high-energy bonds. When an ATP molecule is broken down by an enzyme, the third (terminal) phosphate unit is released as a phosphate group, which is an ion. When this happens, approximately 7.3 kilocalories of energy are released per mole of ATP. (A kilocalorie equals 1,000 calories.) This energy is made available to do the work of the cell.

The adenosine triphosphatase enzyme accomplishes the breakdown of an ATP molecule. The products of ATP breakdown are **adenosine diphosphate (ADP)** and a **phosphate ion.** Adenosine diphosphate and the phosphate ion can be reconstituted to form ATP, much like a battery can be recharged. To accomplish this, synthesis energy must be available. This energy can be made available in the cell through two extremely important processes: photosynthesis (see Chapter 5) and cellular respiration (see Chapter 6).

ATP Production

ATP is generated from ADP and phosphate ions by a complex set of processes occurring in the cell. These processes depend on the activities of a special group of coenzymes. Three important coenzymes are **nicotinamide adenine dinucleotide (NAD), nicotinamide adenine dinucleotide phosphate (NADP),** and **flavin adenine dinucleotide (FAD).**

NAD and NADP are structurally similar to ATP. Both molecules have a nitrogen-containing ring called *nicotinic acid,* which is the chemically active part of the coenzymes. In FAD, the chemically active portion is the flavin group. The vitamin riboflavin is used in the body to produce this flavin group.

All coenzymes perform essentially the same work. During the chemical reactions of metabolism, coenzymes accept electrons and pass them on to other coenzymes or other molecules. The removal of electrons or protons from a coenzyme is *oxidation.* The addition of electrons to a molecule is *reduction.* Therefore, the chemical reactions performed by coenzymes are called *oxidation-reduction reactions.*

The oxidation-reduction reactions performed by the coenzymes and other molecules are essential to the energy metabolism of the cell. Other molecules participating in this energy reaction are called **cytochromes.** Together with the coenzymes, cytochromes accept and release electrons in a system called the *electron transport system.* The passage of energy-rich electrons among cytochromes and coenzymes drains the energy from the electrons to form ATP from ADP and phosphate ions.

The actual formation of ATP molecules requires a complex process involving the creation of a steep proton (hydrogen ion) gradient. This gradient occurs between the membrane-bound compartments of the mitochondria of all cells and the chloroplasts of plant cells. A gradient is formed when large numbers of hydrogen ions are actively transported into the compartment, such as the space between the inner and outer membranes of the mitochondria (called the *inner membrane space*). The protons build up dramatically within the compartment, finally reaching an enormous number. The initial concentration of hydrogen ions was created using the energy released from the electrons as they travel down the electron transport chain. The energy released from the electrons power proton pumps that actively transport hydrogen ions from the matrix of the mitochondria into this inner membrane space.

After large numbers of protons have gathered within the compartments of mitochondria and chloroplasts, they suddenly reverse their directions and diffuse back across the membranes and out of the inner membrane space. The escaping protons release their energy in this motion. This energy is used by enzymes to unite ADP with phosphate ions to form

ATP. The energy is trapped in the high-energy bond of ATP by this process, and the ATP molecules are made available to perform cell work. The process of using the diffusion of hydrogen ions to drive the production of ATP is called **chemiosmosis** (*chemi-* = chemical; *osmosis* = diffusion).

The movement of protons is chemiosmosis because it is a movement of chemicals (in this case, protons) across a semipermeable membrane. Because chemiosmosis occurs in mitochondria and chloroplasts, these organelles play an essential role in the cell's energy metabolism. Chapter 5 explains how energy is trapped in the chloroplasts in plants, while Chapter 6 explains how energy is released in the mitochondria of plant and animal cells.

Chapter Check-Out

Questions

1. The initiation of a chemical reaction requires ______.
   - **a.** an exergonic reaction
   - **b.** energy of activation
   - **c.** a catalyst
2. The rate of enzymatic reactions depends on all of the following conditions EXCEPT ______.
   - **a.** nutrients
   - **b.** temperature
   - **c.** pH
3. True or False: Adenosine triphosphate (ATP) molecules are used to drive chemical reactions.
4. All of the following molecules are types of cofactors used to generate ATP EXCEPT ______.
   - **a.** NAD
   - **b.** FAD
   - **c.** NADH

Answers

1. b
2. a
3. True
4. c

Chapter 5
PHOTOSYNTHESIS

Chapter Check-In

- ❑ Converting sunlight into chemical energy
- ❑ Absorbing light into photosystems to create ATP and NADPH
- ❑ Using ATP and NADPH from the light reactions to make glucose in the Calvin cycle

A great variety of living things on Earth, including all green plants, synthesize their food from carbon dioxide and water. For this to occur, light energy must be transformed into stored chemical energy. This process is called *photosynthesis.*

Figure 5-1 shows the energy transformation in living cells. Light energy is captured in the chloroplasts of plant cells and used to synthesize glucose molecules ($C_6H_{12}O_6$). In the process, oxygen (O_2) is released as a waste product. The glucose is then used to construct cellulose (structural

Figure 5-1 Energy relationships in living cells.

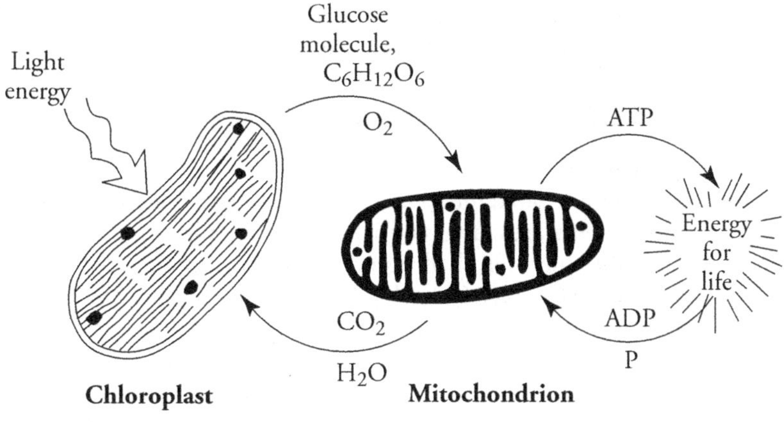

material for the plant), or used immediately to generate ATP for the plant. The mitochondria are able to "burn" glucose in the presence of oxygen, and the energy released is used to fuel the synthesis of ATP from ADP and P (a free, unattached phosphate group). In the reaction, CO_2 and water are released in the mitochondrion to be reused in photosynthesis in the chloroplast. Any excess glucose molecules are linked together in long chains and stored as starch.

The process of photosynthesis has two separate stages: the light reactions and the Calvin cycle (also called the "dark reactions" because the Calvin cycle occurs independent of light). The overall purpose of photosynthesis is to create glucose ($C_6H_{12}O_6$) from carbon dioxide (CO_2). In order for this to occur, the plant needs a significant amount of hydrogen (considering there is a lot of hydrogen in glucose, but none in CO_2) and ATP (the rearrangement and creation of covalent bonds requires a lot of energy). The light reactions are the source for the hydrogen and ATP needed to create glucose in the Calvin cycle.

The Chloroplast

The organelle in which photosynthesis occurs (in the leaves and green stems of plants, for example) is called the **chloroplast.** Chloroplasts are relatively large organelles, containing a watery, protein-rich fluid called *stroma.* The stroma contains many small structures composed of membranes that resemble stacks of coins. Each stack is a *granum* (the plural form is *grana*) and each "coin" in the stack is a hollow **thylakoid.** The light reactions occur on (and within) the thylakoids; the enzymes of the Calvin cycle reside within the stroma. The thylakoids are somewhat similar to the cristae of mitochondria (see Chapter 6).

The Process of Photosynthesis

Light reactions

Pigment molecules organized into photosystems capture sunlight in the chloroplast. **Photosystems** are clusters of light-absorbing pigments with some associated molecules—proton (hydrogen ion) pumps, enzymes, coenzymes, and cytochromes (see Chapter 4). Each photosystem includes a reaction center surrounded by light-harvesting complexes. Each light-harvesting complex contains molecules of a green pigment called **chlorophyll** and molecules of another family of pigments called *carotenoids.* The light-harvesting complex acts as an antenna, focusing the energy of the sun onto the reaction center. The reaction center contains a specific chlorophyll molecule (P680 in photosystem II; P700 in photosystem I). Once the reaction-center chlorophyll collects the energy from the light-harvesting complex, one of its electrons is boosted to a higher energy level. This high-energy electron is then used in the production of either ATP (*photosystem II*) or NADPH (*photosystem I*). Even though photosystem II occurs first in the light-reaction sequence, it was discovered *after* photosystem I (hence the numbering discrepancy).

The light reaction of photosynthesis begins when light is absorbed in photosystem II in the thylakoid membranes. The energy from photons of light captured in the P680 reaction center causes the electrons from P680's chlorophyll to move to a higher, unstable energy level. These electrons pass through a series of cytochromes in the nearby electron-transport system. The energy of the electron is used to pump hydrogen ions from the stroma and across the thylakoid membrane, concentrating the protons inside the thylakoid space. This concentration gradient sets up the potential for chemiosmosis.

The spent electrons from photosystem II enter the P700 reaction center in photosystem I. Sunlight reactivates the electrons, which are again boosted out of the chlorophyll molecules into a higher energy state. The electrons progress through a second electron transport system, but this time there is no proton pumping. Rather, the energized electrons (along with hydrogen ions) reduce $NADP^+$ into NADPH.

Because electrons have flowed out of the P680 reaction center, the chlorophyll molecules are left without a certain number of electrons. Electrons secured from water molecules replace these electrons. Each split water molecule releases two electrons that enter the chlorophyll molecules to replace those lost. The split water molecules also release two hydrogen ions that enter the cytoplasm near the thylakoid and are available to increase the chemiosmotic gradient. The third product of the split water molecules is oxygen. Two oxygen atoms combine with one another to form molecular oxygen (O_2), which is given off as the by-product of photosynthesis; it fills the atmosphere and is used by all oxygen-requiring organisms, including plant and animal cells.

Described below are the *noncyclic light reactions* (see Figure 5-2), the predominant pathway. However, oftentimes the photosynthesizing cell will run out of ATP before it runs out of NADPH; it will then temporarily switch to *cyclic light reactions* in order to focus on creating only ATP. These reactions involve only photosystem I and the P700 reaction center. Excited electrons leave the reaction center, pass through the electron transport pathway that created ATP via chemiosmosis, and follow a special pathway back to P700. Once the ATP concentration is high enough, the cell will switch back to noncyclic photosynthesis, and create both ATP and NADPH.

Figure 5-2 The light reactions of photosynthesis.

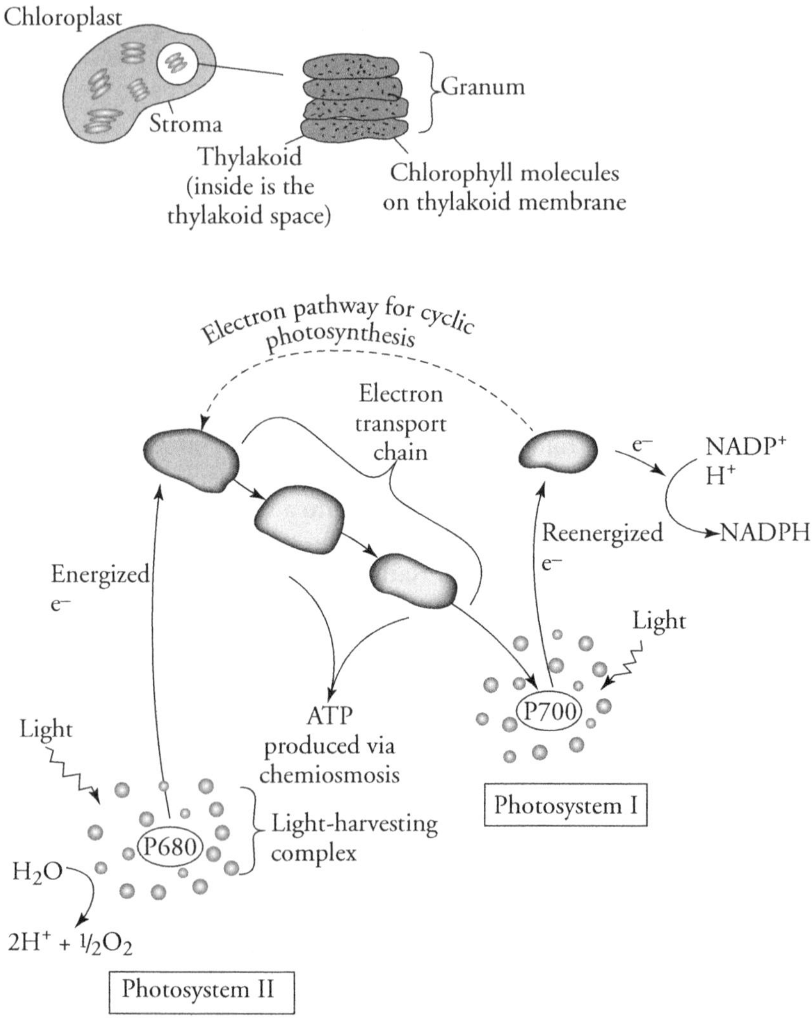

Chemiosmosis

ATP production in the light reactions of photosynthesis occurs by the process of **chemiosmosis** (explained in Chapter 4). Essentially, this process consists of a rush of hydrogen ions across a membrane (the thylakoid membrane, in this case), accompanied by the synthesis of ATP molecules. Biochemists have calculated that the hydrogen ion concentration inside the thylakoid is 10,000 times that in the surrounding stroma. The hydrogen ions will diffuse out of the thylakoid through a special enzyme called ATP synthase. The flow of hydrogen ions powers the enzyme (chemiosmosis), catalyzing the reaction of ADP + P $\rightarrow$ ATP.

Calvin cycle

Glucose and other carbohydrates are synthesized in the carbon-fixing reaction of photosynthesis, often called the *Calvin cycle* after Melvin Calvin, who performed much of the biochemical research (see Figure 5-3). This phase of photosynthesis occurs in the stroma of the plant cell. Though the focus of photosynthesis is the production of glucose, the output of the Calvin cycle is a three-carbon sugar called *glyceraldehyde-3-phosphate*

Figure 5-3 The Calvin cycle.

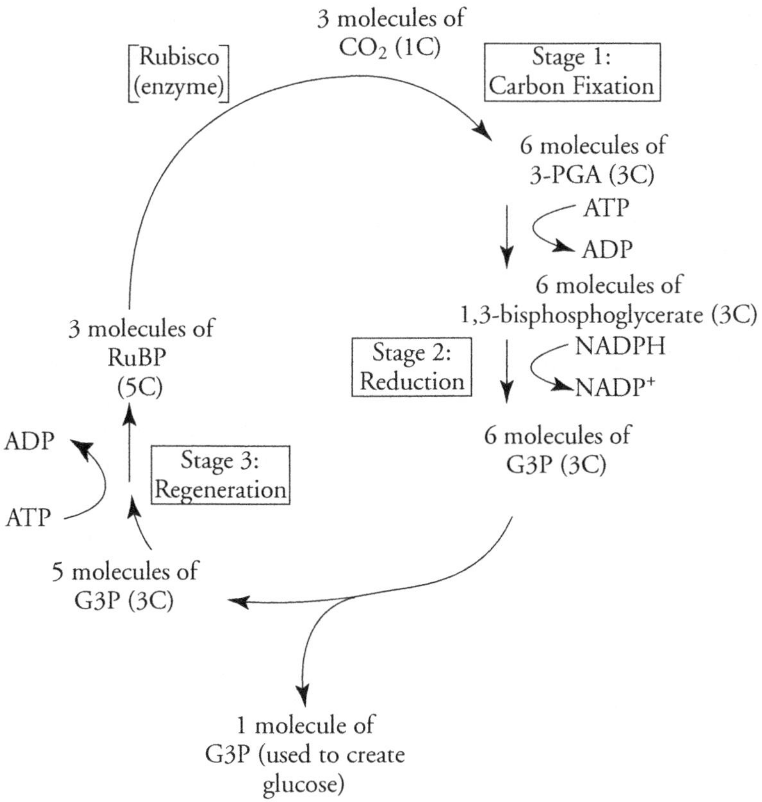

(G3P). The G3P molecules are then used by the cell to create glucose and other organic compounds. The series of reactions that compose the Calvin cycle are complex and intricate. It is easier to consider the three stages of the Calvin cycle: 1) carbon fixation; 2) reduction; and 3) regeneration of RuBP.

Stage 1: Carbon fixation

Carbon fixation refers to the process in which gaseous, atmospheric carbon dioxide is incorporated into organic molecules, thus "fixing" it into a cell. In the Calvin cycle, the carbon dioxide is attached to a five-carbon compound called *ribulose bisphosphate (RuBP)*. The enzyme ribulose bisphosphate carboxylase ("rubisco") catalyzes this reaction. Because the goal is to produce a single G3P molecule (containing three total carbons), there needs to be an input of three CO_2 molecules. RuBP is able to fix one single CO_2; a total of three RuBP are needed to fix three CO_2.

After a carbon dioxide molecule has been joined to each of the three RuBP molecules, an unstable six-carbon product forms. Each of these six-carbon intermediates immediately breaks into two three-carbon molecules, resulting in six total *3-phosphoglycerate* molecules *(3-PGA)*. The atoms of each 3-PGA molecule are rearranged to form six total molecules of 1,3-bisphosphoglycerate. The rearrangement of covalent bonds requires a significant amount of energy in the form of ATP; this ATP was provided back in the light reactions.

Stage 2: Reduction

In order to create G3P (the output of the Calvin cycle), 1,3-bisphosphoglycerate must be "reduced" by adding electrons and hydrogens, provided by the NADPH generated in the light reactions. At this point there are six total molecules of G3P. Only one, however, is "removed" from the Calvin cycle. The remaining five molecules of G3P must continue on in order to regenerate the starting molecule, RuBP.

Stage 3: Regeneration of RuBP

The atoms in the five remaining molecules of G3P are rearranged into three molecules of RuBP (our starting point). In order to do this, the cell must once again provide a significant amount of energy; this energy is ATP created in the light reactions.

All living things on Earth depend on photosynthesis. It is the main mechanism for converting sun energy into chemical energy, making it available for cells to use.

Chapter Check-Out

Questions

1. The coenzyme used to provide a source of hydrogens during the reduction phase of the Calvin cycle is called ______.
   a. NADP
   b. NADH
   c. NADPH

2. Photosynthesis takes place in which of the following organelles?
   a. chlorophyll
   b. chloroplast
   c. stroma

3. The light reactions generate ______, whereas the Calvin cycle produces ______.
   a. ATP and NADPH; CO_2 and water
   b. $C_6H_{12}O_6$; ATP and CO_2
   c. ATP and NADPH; $C_6H_{12}O_6$

4. True or False: The plant cell will temporarily switch to cyclic photosynthesis because it uses more ATP then NADPH in the Calvin cycle.

Answers

1. c
2. b
3. c
4. True

Chapter 6
CELLULAR RESPIRATION

Chapter Check-In

- ❑ Breaking down glucose during glycolysis
- ❑ Moving through the Krebs cycle
- ❑ Transporting electrons and pumping protons
- ❑ Metabolizing glucose by fermentation

Some organisms, such as plants, can trap the energy in sunlight through photosynthesis (see Chapter 5) and store it in the chemical bonds of carbohydrate molecules. The principal carbohydrate formed through photosynthesis is **glucose.** Other types of organisms, such as animals, fungi, many protozoa, and a large portion of bacteria, are unable to perform this process. Therefore, these organisms must rely on the carbohydrates formed in plants to obtain the energy necessary for their metabolic processes.

Animals and other organisms (including plants) release the energy stored in glucose through the process of **cellular respiration.** Glucose is used as long-term energy storage, and large amounts of glucose can be stored as starch or glycogen. When cells need energy to use immediately, a complex series of metabolic processes breaks down the glucose and converts the energy into adenosine triphosphate (ATP) molecules. The ATP can then be used for processes in the cells that require energy, much as a battery powers a mechanical device.

As glucose is broken down through the process of cellular respiration, the by-product carbon dioxide is released from the cell. This carbon dioxide is used during photosynthesis to form new carbohydrates. Furthermore, in the process of cellular respiration, oxygen gas is required to serve as an acceptor of electrons at the end of the electron transport chain. This oxygen is identical to the oxygen gas given off during photosynthesis. Thus, there is an interrelationship between the processes of photosynthesis and cellular respiration. Photosynthesis, using atmospheric carbon dioxide, converts the energy of the sun into a chemical form (glucose) for storage, releasing oxygen as a by-product. Cellular respiration, in turn, "burns" the glucose using oxygen, converting the energy stored in the chemical bonds into ATP.

The overall mechanism of cellular respiration involves four processes: glycolysis, in which glucose molecules are broken down to form pyruvic acid molecules; the Krebs cycle, in which pyruvic acid is further broken down and the energy in the molecules is used to form high-energy compounds, such as nicotinamide adenine dinucleotide (NADH); the electron transport system, in which electrons are transported along a series of coenzymes and cytochromes and the energy in the electrons is released; and chemiosmosis, in which the energy given off by electrons pumps hydrogen ions across a membrane and provides the energy for ATP synthesis. The general chemical equation for cellular respiration is:

$$C_6H_{12}O_6 + 6O_2 \rightarrow 6H_2O + 6CO_2 + \text{energy}$$

Figure 6-1 provides an overview of cellular respiration. Glucose is converted to pyruvic acid in the cytoplasm, which is then used to produce acetyl CoA in the mitochondrion. Finally, the Krebs cycle proceeds in the mitochondrion. Electron transport and chemiosmosis result in energy release; ATP synthesis also occurs in the mitochondrion.

Glycolysis

Glycolysis is the catabolic pathway by which one glucose molecule is broken down into two molecules of pyruvic acid (also called pyruvate), releasing energy in the process. Glycolysis is the earliest method of energy production, utilized by the first prokaryotic cells present on Earth. There was no atmospheric oxygen present during this time; thus, glycolysis is an anaerobic process that evolved to occur in the cytoplasm of cells (since prokaryotic cells do not contain any organelles). It is the most widespread energy-producing pathway on Earth.

The process of glycolysis consists of ten enzyme-catalyzed steps. In the first and third steps of the pathway, ATP energizes the molecules. Thus, two ATP molecules must be expended in the process. Further along in the process, the six-carbon glucose molecule converts into intermediary compounds and is then split into two three-carbon compounds. The latter undergo additional conversions and eventually form pyruvic acid at the conclusion of the process.

Figure 6-1 An overview of cellular respiration.

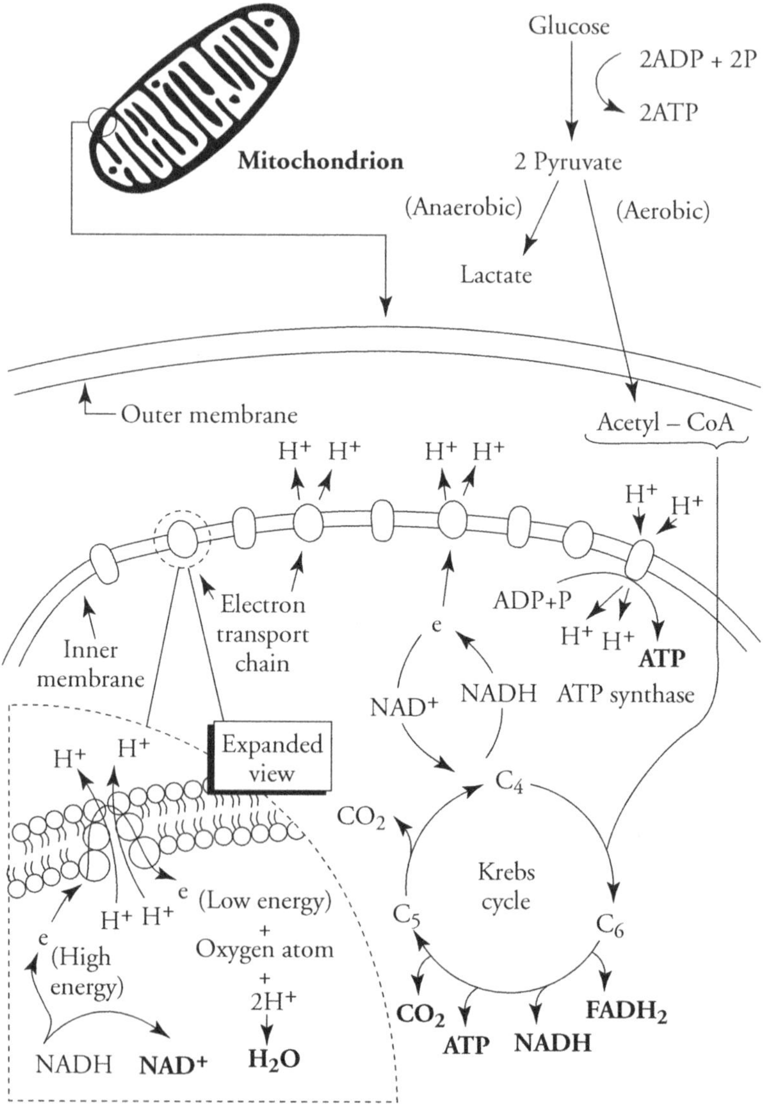

During the latter stages of glycolysis, four ATP molecules are synthesized using the energy given off during the chemical reactions. Thus, four ATP molecules are synthesized, and two ATP molecules are used during glycolysis, for a net gain of two ATP molecules.

Another reaction during glycolysis yields enough energy to convert NAD to NADH (plus a hydrogen ion). The reduced coenzyme (NADH) will later be used in the electron transport system, and its energy will be released. During glycolysis, two NADH molecules are produced.

Because glycolysis does not require oxygen, the process is considered to be anaerobic. For certain anaerobic organisms, such as some bacteria and fermentation yeasts, glycolysis is the sole source of energy.

Glycolysis is a somewhat inefficient process because much of the cellular energy remains in the two molecules of pyruvic acid that are created. Interestingly, this process is somewhat similar to a reversal of photosynthesis (see Chapter 5).

Krebs Cycle

Following glycolysis, the mechanism of cellular respiration involves another multi-step process—the Krebs cycle, which is also called the citric acid cycle or the tricarboxylic acid cycle. The **Krebs cycle** uses the two molecules of pyruvic acid formed in glycolysis and yields high-energy molecules of NADH and flavin adenine dinucleotide ($FADH_2$), as well as some ATP.

The Krebs cycle occurs in the mitochondrion of a cell (see Figure 6-1). This sausage-shaped organelle possesses inner and outer membranes and, therefore, inner and outer compartments. The inner membrane is folded over itself many times; the folds are called *cristae.* They are somewhat similar to the thylakoid membranes in chloroplasts (see Chapter 5). Located along the cristae are the important enzymes necessary for the proton pump and for ATP production.

Prior to entering the Krebs cycle, the pyruvic acid molecules are altered. Each three-carbon pyruvic acid molecule undergoes conversion to a substance called acetyl-coenzyme A, or acetyl-CoA. During the process, the pyruvic acid molecule is broken down by an enzyme, one carbon atom is released in the form of carbon dioxide, and the remaining two carbon atoms are combined with a coenzyme called coenzyme A. This combination forms acetyl-CoA. In the process, electrons and a hydrogen ion are transferred to NAD to form high-energy NADH.

Acetyl-CoA enters the Krebs cycle by combining with a four-carbon acid called oxaloacetic acid. The combination forms the six-carbon acid called citric acid. Citric acid undergoes a series of enzyme-catalyzed conversions. The conversions, which involve up to ten chemical reactions, are all brought about by enzymes. In many of the steps, high-energy electrons are released to NAD. The NAD molecule also acquires a hydrogen ion and becomes NADH. In one of the steps, FAD serves as the electron acceptor, and it acquires two hydrogen ions to become $FADH_2$. Also, in one of the reactions, enough energy is released to synthesize a molecule of ATP. Because for each glucose molecule there are two pyruvic acid molecules entering the system, two ATP molecules are formed.

Also during the Krebs cycle, the two carbon atoms of acetyl-CoA are released, and each forms a carbon dioxide molecule. Thus, for each acetyl-CoA molecule entering the cycle, two carbon dioxide molecules are formed. Two acetyl-CoA molecules enter the cycle, and each has two carbon atoms, so four carbon dioxide molecules will form. Add these four molecules to the two carbon dioxide molecules formed in the conversion of pyruvic acid to acetyl-CoA, and it adds up to six carbon dioxide molecules. These six CO_2 molecules are given off as waste gas in the Krebs cycle. They represent the six carbons of glucose that originally entered the process of glycolysis.

At the end of the Krebs cycle, the final product is oxaloacetic acid. This is identical to the oxaloacetic acid that begins the cycle. Now the molecule is ready to accept another acetyl-CoA molecule to begin another turn of the cycle. All told, the Krebs cycle forms (per two molecules of pyruvic acid) two ATP molecules, ten NADH molecules, and two $FADH_2$ molecules. The NADH and the $FADH_2$ will be used in the electron transport system.

Electron Transport System

The electron transport system occurs in the cristae of the mitochondria, where a series of cytochromes (enzymes) and coenzymes exist. These cytochromes and coenzymes act as carrier molecules and transfer molecules. They accept high-energy electrons and pass the electrons to the next molecule in the system. At key proton-pumping sites, the energy of the electrons transports protons across the membrane into the outer compartment of the mitochondrion.

Each NADH molecule is highly energetic, which accounts for the transfer of six protons into the outer compartment of the mitochondrion. Each $FADH_2$ molecule accounts for the transfer of four protons. The flow of electrons is similar to that taking place in photosynthesis. Electrons pass from NAD to FAD, to other cytochromes and coenzymes, and eventually they lose much of their energy. In cellular respiration, the final electron acceptor is an oxygen atom. In their energy-depleted condition, the electrons unite with an oxygen atom. The electron-oxygen combination then reacts with two hydrogen ions (protons) to form a water molecule (H_2O).

The role of oxygen in cellular respiration is substantial. As a final electron acceptor, it is responsible for removing electrons from the electron transport system. If oxygen were not available, electrons could not be passed among the coenzymes, the energy in electrons could not be released, the proton pump could not be established, and ATP could not be produced. In humans, breathing is the essential process that brings oxygen into the body for delivery to the cells to participate in cellular respiration.

Chemiosmosis

The actual production of ATP in cellular respiration takes place through the process of chemiosmosis (see Chapter 4). **Chemiosmosis** involves the pumping of protons through special channels in the membranes of mitochondria from the inner to the outer compartment. The pumping establishes a proton (H^+) gradient. After the gradient is established, protons diffuse down the gradient through a transport protein called ATP synthase. The flow of hydrogens catalyzes the pairing of a phosphate with ADP, forming ATP.

The energy production of cellular respiration is substantial. Most biochemists agree that 36 molecules of ATP can be produced for each glucose molecule during cellular respiration as a result of the Krebs cycle reactions, the electron transport system, and chemiosmosis. Also, two ATP molecules are produced through glycolysis, so the net yield is 38 molecules of ATP. These ATP molecules may then be used in the cell for its needs. However, the ATP molecules cannot be stored for long periods of time, so cellular respiration must constantly continue in order to regenerate the ATP molecules as they are used. Each ATP molecule is capable of releasing 7.3 kilocalories of energy per mole.

Fermentation

Fermentation is an anaerobic process in which energy can be released from glucose even though oxygen is not available. Fermentation occurs in yeast cells, and a form of fermentation takes place in bacteria and in the muscle cells of animals.

In yeast cells, glucose can be metabolized through cellular respiration as in other cells. When oxygen is lacking, however, glucose is still metabolized to pyruvic acid via glycolysis. The pyruvic acid is converted first to acetaldehyde and then to ethyl alcohol. The net gain of ATP to the yeast cell is two molecules—the two molecules of ATP normally produced in glycolysis.

Yeasts are able to participate in fermentation because they have the necessary enzyme to convert pyruvic acid to ethyl alcohol. This process is essential because it removes electrons and hydrogen ions from NADH during glycolysis. The effect is to free the NAD so it can participate in future reactions of glycolysis. The net gain to the yeast cell of two ATP molecules permits it to remain alive for some time. However, when the percentage of ethyl alcohol reaches approximately 15 percent, the alcohol kills the yeast cells.

Yeast is used in both bread and alcohol production. Alcohol fermentation is the process that yields beer, wine, and other spirits. The carbon dioxide given off during fermentation supplements the carbon dioxide given off during the Krebs cycle and causes bread to rise.

In muscle cells, another form of fermentation takes place. When muscle cells contract too frequently (as in strenuous exercise), they rapidly use up their oxygen supply. As a result, the electron transport system and Krebs cycle slow considerably, and ATP production is slowed. However, muscle cells have the ability to produce a small amount of ATP through glycolysis in the absence of oxygen. The muscle cells convert glucose to pyruvic acid. An enzyme in the muscle cells then converts the pyruvic acid to lactic acid. As in the yeast, this reaction frees up the NAD while providing the cells with two ATP molecules from glycolysis. Eventually, however, the lactic acid buildup causes intense fatigue, and the muscle stops contracting.

Chapter Check-Out

Questions

1. True or False: Cellular respiration takes place in chloroplasts.
2. Prior to entering the Krebs cycle, two molecules of ______ are broken down to form ______.
   a. $C_6H_{12}O_6$; pyruvate
   b. $C_6H_{12}O_6$; acetyl-CoA
   c. pyruvate; acetyl-CoA
   d. acetyl-CoA; $C_6H_{12}O_6$

3. For one molecule of glucose, ______ molecule(s) of ATP is/are formed during the Krebs cycle.
   a. one
   b. two
   c. four
   d. eight
4. For each glucose molecule, ______ molecules of ATP are formed during respiration.
   a. 12
   b. 24
   c. 36
   d. 38

Answers

1. False
2. c
3. b
4. d

Chapter 7

MITOSIS AND CELL REPRODUCTION

Chapter Check-In

- ❑ Reproducing within the nucleus
- ❑ Cycling through cell division
- ❑ Separating nuclear components

A distinguishing feature of a living thing is that it reproduces independently of other living things. This reproduction occurs at the cellular level. In certain parts of the body, such as along the gastrointestinal tract, the cells reproduce often. In other parts of the body, such as in the nervous system, the cells reproduce less frequently. With the exception of only a few kinds of cells, such as red blood cells (which lack nuclei when fully mature), all cells of the human body reproduce.

Cell Nucleus

In eukaryotic cells (see Chapter 3), the structure and contents of the nucleus are of fundamental importance to an understanding of cell reproduction. The nucleus contains the hereditary material (DNA) of the cell assembled into chromosomes. In addition, the nucleus usually contains one or more prominent **nucleoli** (dense bodies that are the site of ribosome synthesis).

The nucleus is surrounded by a nuclear envelope consisting of a double membrane that is continuous with the endoplasmic reticulum. Transport of molecules between the nucleus and cytoplasm is accomplished through a series of nuclear pores lined with proteins that facilitate the passage of molecules out of the nucleus. The proteins provide a certain measure of selectivity in the passage of molecules across the nuclear membrane.

The nuclear material consists of deoxyribonucleic acid (DNA) organized into long strands. The strands of DNA are composed of nucleotides bonded to one another by covalent bonds. DNA molecules are extremely long relative to the cell; there are approximately 6 feet of DNA in a single human cell. However, in the chromosome, the DNA is condensed and packaged with protein into manageable bodies. The mass of DNA material and its associated protein is **chromatin.**

To form chromatin, the DNA molecule is wound around globules of a protein called **histone.** The units formed in this way are *nucleosomes.* Millions of nucleosomes are connected by short stretches of histone protein, much like beads on a string. The configuration of the nucleosomes in a coil causes additional coiling of the DNA and the eventual formation of the chromosome.

Cell Cycle

The **cell cycle** involves many repetitions of cellular growth and reproduction. With few exceptions (for example, red blood cells), all the cells of living things undergo a cell cycle.

The cell cycle is generally divided into two phases: interphase and mitosis. During **interphase,** the cell spends most of its time performing the functions that make it unique. **Mitosis** is the phase of the cell cycle during which the cell divides into two daughter cells.

Interphase

The interphase stage of the cell cycle includes three distinctive parts: the G_1 phase, the S phase, and the G_2 phase. The **G_1 phase** follows mitosis and is the period in which the cell is synthesizing its structural proteins and enzymes to perform its functions. For example, a pancreas cell in the G_1 phase will produce and secrete insulin, a muscle cell will undergo the contractions that permit movement, and a salivary gland cell will secrete salivary enzymes to assist digestion. During the G_1 phase, each chromosome consists of a single molecule of DNA and its associated histone protein. In normal human cells, there are 46 chromosomes per cell (except in sex cells with 23 chromosomes and red blood cells with no nucleus and, hence, no chromosomes).

During the **S phase** of the cell cycle, the DNA within the nucleus replicates. During this process, each chromosome is faithfully copied, so by the end of the S phase, two DNA molecules exist for each one formerly present in the G_1 phase. Human cells contain 92 chromosomes per cell in the S phase.

In the **G_2 phase,** the cell prepares for mitosis. Proteins organize themselves to form a series of fibers called the *spindle,* which is involved in chromosome movement during mitosis. The spindle is constructed from amino acids for each mitosis and then taken apart at the conclusion of the process. Spindle fibers are composed of microtubules.

Mitosis

The term *mitosis* is derived from the Latin stem *mito,* meaning "threads." When mitosis was first described a century ago, scientists had seen "threads" within cells, so they gave the name "mitosis" to the process of "thread movement." During mitosis, the nuclear material becomes visible as threadlike chromosomes. The chromosomes organize in the center of the cell and then they separate, and 46 chromosomes move into each new cell that forms.

Mitosis is a continuous process, but for convenience in denoting which portion of the process is taking place, scientists divide mitosis into a series of phases: prophase, metaphase, anaphase, and telophase (see Figure 7-1):

- **Prophase:** Mitosis begins with the condensing of the chromatin to form chromosomes in the phase called prophase. Two copies of each chromosome exist; each one is a **chromatid.** Two chromatids are joined to one another at a region called the **centromere.** As prophase unfolds, the chromatids become visible in pairs (called sister chromatids), the spindle fibers form, the nucleoli disappear, and the nuclear envelope dissolves.

  In animal cells during prophase, microscopic bodies called **centrioles** begin to migrate to opposite sides of the cell. When the centrioles reach the poles of the cell, they produce and are then surrounded by a series of radiating microtubules called an *aster.* Centrioles and asters are not present in most plant or fungal cells.

  As prophase continues, the chromatids attach to spindle fibers that extend out from opposite poles of the cell. The spindle fibers attach at the region of the centromere at a structure called the **kinetochore,** an area of protein in the centromere region. Eventually, all pairs of chromatids reach the center of the cell, a region called the *equatorial plate.*

Figure 7-1 The process of mitosis, in which the chromosomes of a cell duplicate and pass into two daughter cells.

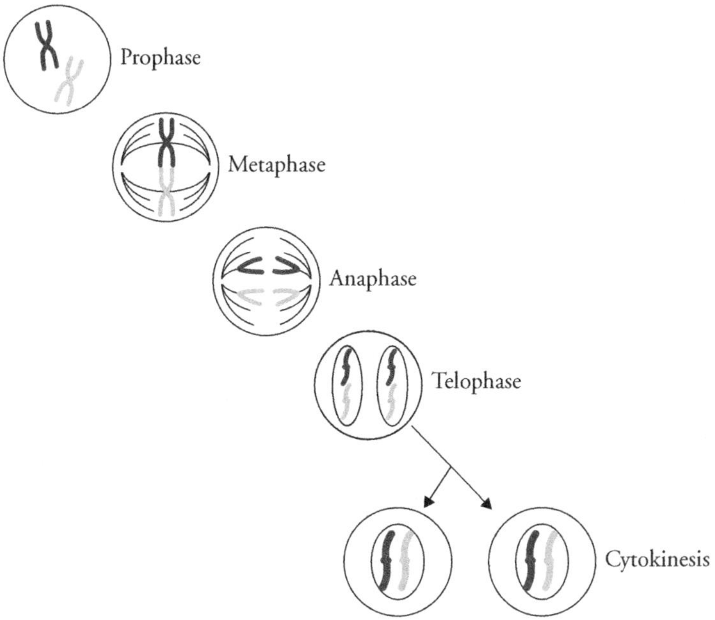

- **Metaphase:** Metaphase is the stage of mitosis in which the pairs of chromatids line up on the equatorial plate. This region is also called the metaphase plate. In a human cell, 92 chromosomes in 46 pairs align at the equatorial plate. Each pair is connected at the centromere, where the spindle fiber is attached (more specifically at the kinetochore).
- **Anaphase:** At the beginning of anaphase, the sister chromatids move apart from one another. The chromatids are called **chromosomes** after the separation. Each chromosome is attached to a spindle fiber, and the members of each chromosome pair are drawn to opposite poles of the cell by the spindle fibers. During anaphase, the chromosomes can be seen moving. They take on a rough V shape because of their midregion attachment to the spindle fibers. The movement toward the poles is accomplished by several mechanisms, such as an elongation of the spindle fibers, which results in pushing the poles apart.

  The result of anaphase is an equal separation and distribution of the chromosomes. In human cells, a total of 46 chromosomes move to each pole as the process of mitosis continues.

- **Telophase:** In telophase, the chromosomes finally arrive at the opposite poles of the cell. The distinct chromosomes begin to fade from sight as masses of chromatin are formed again. The events of telophase are essentially the reverse of those in prophase. The spindle is dismantled, its amino acids are recycled, the nucleoli reappear, and the nuclear envelope is reformed.
- **Cytokinesis:** Cytokinesis is the process in which the cytoplasm divides and two separate cells form. Note that cytokinesis is separate from the four stages of mitosis. In animal cells, cytokinesis begins with the formation of a cleavage furrow in the center of the cell. With the formation of the furrow, the cell membrane begins to pinch into the cytoplasm, and the formation of two cells begins. This process is often referred to as *cell cleavage.* Microfilaments contract during cleavage and assist the division of the cell into two daughter cells.

  In plant cells, cytokinesis occurs by a different process because a rigid cell wall is involved. Cleavage does not take place in plant cells. Rather, a new cell wall is assembled at the center of the cell, beginning with vesicles formed from the Golgi apparatus (see Chapter 3). As the vesicles join, they form a double membrane called the *cell plate.* The cell plate forms in the middle of the cytoplasm and grows outward to fuse with the cell membrane. The cell plate separates the two daughter cells. As cell wall material is laid down, the two cells move apart from one another to yield two new daughter cells.

Mitosis serves several functions in living cells. In many simple organisms (such as fungi and bacteria), it is the method for asexual reproduction. In multicellular organisms, mitosis allows the entire organism to grow by forming new cells and replacing older cells. Once an egg cell is fertilized by a sperm cell, the resulting diploid zygote then has two sets of chromosomes, as supplied by the parent cells. Through the process of mitosis, the zygote is able to divide successively to produce many cells, each with identical DNA. As the organism develops, the cells begin to differentiate into distinct tissues and organs, and the differing functions of the complex organism take shape.

Chapter Check-Out

Questions

1. Red blood cells contain ______ chromosomes.
   a. 0
   b. 23
   c. 46
2. Mitosis is divided into all of the following phases EXCEPT ______.
   a. interphase
   b. prophase
   c. metaphase
3. ______ attach to a region of the centromere called the ______.
   a. Spindle fibers; kinetochore
   b. Centriole; kinetochore
   c. Asters; centriole

Answers

1. a
2. a
3. a

Chapter 8

MEIOSIS AND GAMETE FORMATION

Chapter Check-In

- ❑ Generating haploid cells
- ❑ Dividing by meiosis
- ❑ Crossing over

Most plant and animal cells are diploid. The term **diploid** is derived from the Greek *diplos,* meaning "double" or "two"; the term implies that the cells of plants and animals have pairs of chromosomes. In human cells, for example, 46 chromosomes are organized in 23 pairs. Hence, human cells are diploid in that they have a pair of 23 individual chromosomes.

During sexual reproduction, the sex cells of parent organisms unite with one another and form a fertilized egg cell **(zygote).** In this situation, each sex cell is a **gamete.** The gametes of human cells are **haploid,** from the Greek *haplos,* meaning "single." This term implies that each gamete contains half of the 46 chromosomes—23 chromosomes in humans. When the human gametes unite with one another, the original diploid condition of 46 chromosomes is reestablished. Mitosis then brings about the development of the diploid cell into a multicellular organism.

The process by which the chromosome number is halved during gamete formation is **meiosis.** In meiosis, a cell containing the diploid number of chromosomes is converted into four cells, each having the haploid number of chromosomes. In human cells undergoing meiosis, for example, a cell containing 46 chromosomes yields four cells, each with 23 chromosomes.

Meiosis occurs by a series of steps that resemble the steps of mitosis. Two major phases of meiosis occur: meiosis I and meiosis II. During meiosis I, a single cell divides into two. During meiosis II, those two cells each divide again. The same demarcating phases of mitosis take place in meiosis I and meiosis II—prophase, metaphase, anaphase, and telophase—but with some variations contained therein.

As shown in Figure 8-1, first, the chromosomes of a cell are divided into two cells. The chromosomes of the two cells then separate and pass into four daughter cells. The parent cell is diploid, while each of the daughter cells has a single set of chromosomes and is haploid. Synapsis and crossing over occur in the prophase I stage.

The members of each chromosome pair within a cell are called *homologous chromosomes.* Homologous chromosomes are similar but not identical. They may carry different versions of the same genetic information. For example, one homologous chromosome may carry the information for blond hair while the other homologous chromosome may carry the information for black hair.

Meiosis

As a cell prepares to enter meiosis, each of its chromosomes has duplicated in the synthesis stage (S) of the cell cycle, as in mitosis. Each chromosome thus consists of two sister chromatids.

Meiosis I

At the beginning of meiosis I, a human cell contains 46 chromosomes, or 92 chromatids (the same number as during mitosis). Meiosis I proceeds through the following phases:

- **Prophase I:** Prophase I is similar in some ways to prophase in mitosis. The chromatids shorten and thicken and become visible under a microscope. An important difference, however, is that a process called **synapsis** occurs. Synapsis is when the homologous chromosomes migrate toward one another and join to form a *tetrad* (the combination of four chromatids, two from each homologous chromosome). A second process called **crossing over** also takes place during prophase I. In this process, segments of DNA from one chromatid in the tetrad pass to another chromatid in the tetrad. These exchanges of chromosomal segments occur in a complex and poorly understood manner. They result in a genetically new chromatid.

Figure 8-1 The process of meiosis, in which four haploid cells are formed.

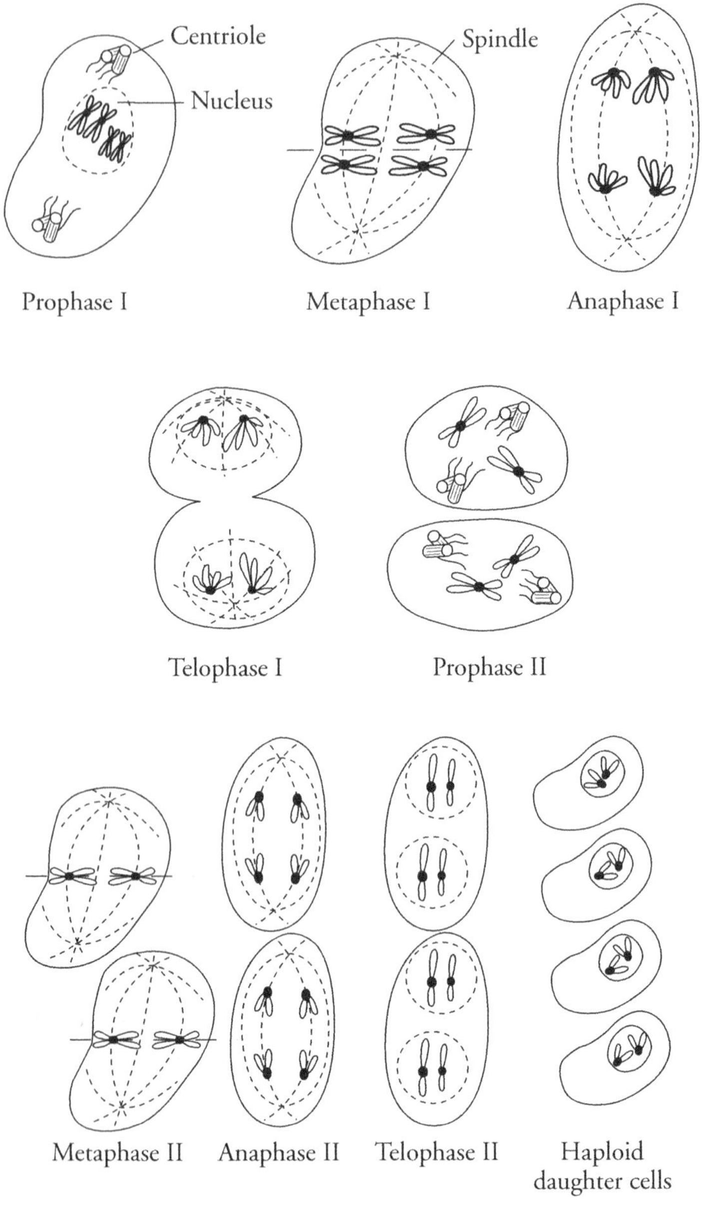

Crossing over is an important driving force of evolution. After crossing over has taken place, the homologous pair of chromosomes is genetically different.

- **Metaphase I:** In metaphase I of meiosis, the tetrads align on the equatorial plate (as in mitosis). The centromeres attach to spindle fibers, which extend from the poles of the cell. One centromere attaches per spindle fiber.
- **Anaphase I:** In anaphase I, the homologous chromosomes or tetrads separate. One homologous chromosome (consisting of two chromatids) moves to one side of the cell, while the other homologous chromosome (consisting of two chromatids) moves to the other side of the cell. The result is that 23 chromosomes (each consisting of two chromatids) move to one pole, and 23 chromosomes (each consisting of two chromatids) move to the other pole. Essentially, the chromosome number of the cell is halved once meiosis I is completed. For this reason the process is a reduction-division.
- **Telophase I:** In telophase I of meiosis, the nucleus reorganizes, the chromosomes become chromatin, and the cell membrane begins to pinch inward. Cytokinesis occurs immediately following telophase I. This process occurs differently in plant and animal cells, just as in mitosis.

Meiosis II

Meiosis II is the second major subdivision of meiosis. It occurs in essentially the same way as mitosis. In meiosis II, a cell contains a single set of chromosomes. Each chromosome, however, still has its duplicated sister chromatid attached. Meiosis II segregates the sister chromatids into separate cells. Meiosis II proceeds through the following phases:

- **Prophase II:** Prophase II is similar to the prophase of mitosis. The chromatin material condenses, and each chromosome contains two chromatids attached by the centromere. The 23 chromatid pairs, a total of 46 chromatids, then move to the equatorial plate.
- **Metaphase II:** In metaphase II of meiosis, the 23 chromatid pairs gather at the center of the cell prior to separation. This process is identical to metaphase in mitosis, except that this is occurring in a haploid versus a diploid cell.

- **Anaphase II:** During anaphase II of meiosis, the centromeres divide and sister chromatids separate, at which time they are referred to as non-replicated chromosomes. Spindle fibers move chromosomes to each pole. In all, 23 chromosomes move to each pole. The forces and attachments that operate in mitosis also operate in anaphase II.
- **Telophase II:** During telophase II, the chromosomes gather at the poles of the cells and become indistinct. Again, they form a mass of chromatin. The nuclear envelope develops, the nucleoli reappear, and the cells undergo cytokinesis.

During meiosis II, each cell containing 46 chromatids yields two cells, each with 23 chromosomes. Originally, there were two cells that underwent meiosis II; therefore, the result of meiosis II is four cells, each with 23 chromosomes. Each of the four cells is haploid; that is, each cell contains a single set of chromosomes.

The 23 chromosomes in the four cells from meiosis are not identical because crossing over has taken place in prophase I. The crossing over yields genetic variation so that each of the four resulting cells from meiosis differs from the other three. Thus, meiosis provides a mechanism for producing variations in the chromosomes. Also, it accounts for the formation of four haploid cells from a single diploid cell.

Meiosis in Humans

In humans, meiosis is the process by which sperm cells and egg cells are produced. In the male, meiosis takes place after puberty. Diploid cells within the testes undergo meiosis to produce haploid **sperm cells** with 23 chromosomes. A single diploid cell yields four haploid sperm cells through meiosis.

In females, meiosis begins during the fetal stage when a series of diploid cells enter meiosis I. At the conclusion of meiosis I, the process comes to a halt, and the cells gather in the ovaries. At puberty, meiosis resumes. One cell at the end of meiosis I enters meiosis II each month. The result of meiosis II is a single **egg cell** per cycle (the other meiotic cells disintegrate). Each egg cell contains 23 chromosomes and is haploid.

The union of the egg cell and the sperm cell leads to the formation of a fertilized egg cell with 46 chromosomes, or 23 pairs. Fertilization restores the diploid number of chromosomes. The fertilized egg cell, a diploid, is a zygote. Further divisions of the zygote by mitosis eventually yield a complete human being.

Chapter Check-Out

Questions

1. As a cell enters meiosis I, the cell contains ______ chromosomes, or ______ chromatids.
   a. 46; 92
   b. 23; 46
   c. 46; 23
   d. 92; 46
2. Each diploid human cell contains ______ pairs of homologous chromosomes.
   a. 23
   b. 36
   c. 46
   d. 96
3. A reproductive cell containing 46 chromosomes will yield ______ cells with ______ chromosomes each after meiosis.
   a. 2; 23
   b. 4; 23
   c. 4; 46
   d. 8; 23
4. True or False: As a cell exits meiosis I and prepares to enter meiosis II, the cell reproduces its DNA.

Answers

1. a
2. a
3. b
4. False

Chapter 9

CLASSICAL (MENDELIAN) GENETICS

Chapter Check-In

- ❑ Introducing Gregor Mendel
- ❑ Predicting inheritance patterns
- ❑ Defining the principles of genetics

Genetics is the study of how genes bring about characteristics, or traits, in living things and how those characteristics are inherited. Genes are specific sequences of nucleotides that code for particular proteins. Through the processes of meiosis and sexual reproduction, genes are transmitted from one generation to the next.

Augustinian monk Gregor Mendel developed the science of genetics. Mendel performed his experiments in the 1860s and 1870s, but the scientific community did not accept his work until early in the twentieth century. Because the principles established by Mendel form the basis for genetics, the science is often referred to as *Mendelian genetics.* It is also called *classical genetics* to distinguish it from another branch of biology known as molecular genetics (see Chapter 10).

Mendel believed that factors pass from parents to their offspring, but he did not know of the existence of DNA. Modern scientists accept that **genes** are composed of segments of DNA molecules that control discrete hereditary characteristics.

Most complex organisms have cells that are diploid. **Diploid** cells have a double set of chromosomes, one from each parent. For example, human cells have a double set of chromosomes consisting of 23 pairs, or a total of 46 chromosomes. In a diploid cell, there are two genes for each characteristic. In preparation for sexual reproduction, the diploid number of chromosomes is reduced to a haploid number. That is, diploid cells are reduced to cells that have a single set of chromosomes. These haploid cells are **gametes,** or sex cells, and they are formed through meiosis (see Chapter 8). When gametes come together in sexual reproduction, the diploid condition is reestablished.

The offspring of sexual reproduction obtain one gene of each type from each parent. The different forms of a gene are called **alleles.** In humans, for example, there are two alleles for earlobe construction. One allele is for earlobes that are attached, while the other allele is for earlobes that hang free. The type of earlobe a person has is determined by the alleles inherited from the parents.

The set of all genes that specify an organism's traits is known as the organism's **genome.** The genome for a human cell consists of about 20,000 genes. The gene composition of a living organism is its **genotype.** For a person's earlobe shape, the genotype may consist of two alleles for attached earlobes, or two alleles for free earlobes, or one allele for attached earlobes and one allele for free earlobes.

The expression of the genes is referred to as the **phenotype** of a living thing. If a person has attached earlobes, the phenotype is "attached earlobes." If the person has free earlobes, the phenotype is "free earlobes." Even though three genotypes for earlobe shape are possible, only two phenotypes (attached earlobes and free earlobes) are possible.

The two paired alleles in an organism's genotype may be identical, or they may be different. An organism's condition is said to be **homozygous** when two identical alleles are present for a particular characteristic. In contrast, the condition is said to be **heterozygous** when two different alleles are present for a particular characteristic. In a homozygous individual, the alleles express themselves. In a heterozygous individual, the alleles may interact with one another, and in many cases, only one allele is expressed.

When one allele expresses itself and the other does not, the one expressing itself is the **dominant allele.** The "overshadowed" allele is the **recessive allele.** In humans, the allele for free earlobes is the dominant allele. If this allele is present with the allele for attached earlobes, the allele for free earlobes expresses itself, and the phenotype of the individual is "free earlobes." Dominant alleles always express themselves, while recessive alleles express themselves only when two recessive alleles exist together in an individual. Thus, a person having free earlobes can have one dominant allele or two dominant alleles, while a person having attached earlobes must have two recessive alleles.

Inheritance Patterns

Mendel was the first scientist to develop a method for predicting the outcome of inheritance patterns. He performed his work with pea plants, studying seven traits: plant height, pod shape, pod color, seed shape, seed color, flower color, and flower location. Pea plants pollinate themselves. Therefore, over many generations, pea plants develop individuals that are homozygous for particular characteristics. These populations are known as *pure lines.*

In his work, Mendel took pure-line pea plants and cross-pollinated them with other pure-line pea plants. He called these plants the *parent generation.* When Mendel crossed pure-line tall plants with pure-line short plants, he discovered that all the plants resulting from this cross were tall. He called this generation the *F_1 generation* (first filial generation). Next, Mendel crossed the offspring of the F_1 generation tall plants among themselves to produce a new generation called the *F_2 generation* (second filial generation). Among the plants in this generation, Mendel observed that three-fourths of the plants were tall and one-fourth of the plants were short.

Mendel's laws of genetics

Mendel conducted similar experiments with the other pea plant traits. Over many years, he formulated several principles that are known today as Mendel's laws of genetics. His laws include the following:

1. **Mendel's law of dominance:** When an organism has two different alleles for a trait, one allele dominates.
2. **Mendel's law of segregation:** During gamete formation by a diploid organism, the pair of alleles for a particular trait separate, or segregate, during the formation of gametes (as in meiosis).
3. **Mendel's law of independent assortment:** The members of a gene pair separate from one another independent of the members of other gene pairs. (These separations occur in the formation of gametes during meiosis.)

Mendelian crosses

An advantage of genetics is that scientists can predict the probability of inherited traits in offspring by performing a genetic cross (also called a *Mendelian cross*). To predict the possibility of an individual trait, several steps are followed. First, a symbol is designated for each allele in the gene pair. The dominant allele is represented by a capital letter and the recessive allele by the corresponding lowercase letter, such as *E* for free earlobes and *e* for attached earlobes. For a homozygous dominant individual, the genotype would be *EE;* for a heterozygous individual, the genotype would be *Ee;* and for a homozygous recessive individual, the genotype would be *ee.*

The next step in performing a genetic cross is determining the genotypes of the parents and the genotype of the gametes. A heterozygous male and a heterozygous female to be crossed have the genotypes of *Ee* and *Ee.* During meiosis, the allele pairs separate. A sperm cell contains either an *E* or an *e,* while the egg cell also contains either an *E* or an *e.*

To continue the genetics problem, a Punnett square is used. A *Punnett square* is a boxed figure used to determine the probability of genotypes and phenotypes in the offspring of a genetic cross. The possible gametes produced by the female are indicated at the top of the square, while the possible gametes produced by the male are indicated at the left side of the square. Figure 9-1 shows the Punnett square for the earlobe example.

Figure 9-1 An example of a Punnett square.

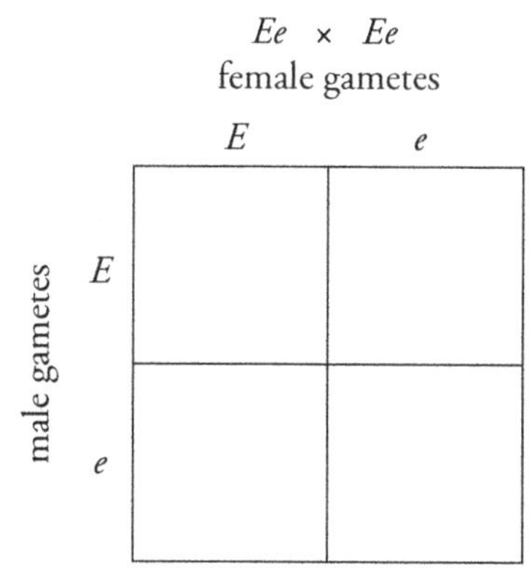

Continuing, all of the possible combinations of alleles are considered. This is done by filling in each square with the alleles above it and at its left. This is done as shown in Figure 9-2.

From the Punnett square, the phenotype of each possible genotype can be determined. For example, the offspring having *EE, Ee,* and *Ee* will have free earlobes. Only the offspring with the genotype *ee* will have attached earlobes. Therefore, the ratio of phenotypes is three with free earlobes to one with attached earlobes (3:1). The ratio of genotypes is 1:2:1 (1 *EE* : 2 *Ee* : 1 *ee*).

Figure 9-2 The Punnett square is used to determine the probabilities of the genotypes and phenotypes in the offspring of a genetic cross.

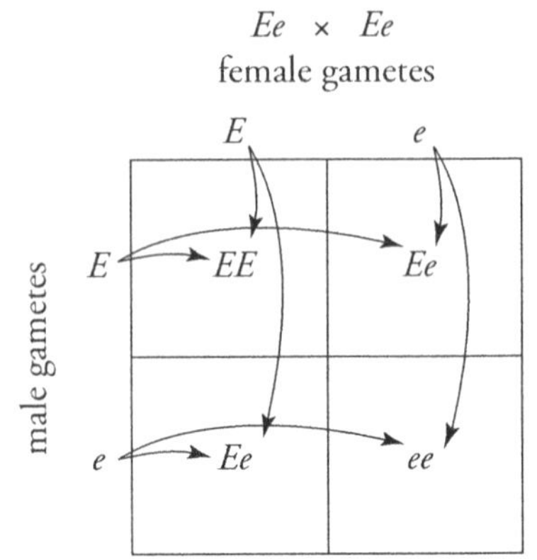

Phenotypic ratio of offspring—
3 free : 1 attached

Genotypic ratio of offspring—
1 *EE* : 2 *Ee* : 1 *ee*

Principles of Genetics

Mendel's studies have provided scientists with the basis for mathematically predicting the probabilities of genotypes and phenotypes in the offspring of a genetic cross. But not all genetic observations can be explained and predicted based on Mendelian genetics. Other complex and distinct genetic phenomena may also occur. Several complex genetic concepts, described in this section, explain such distinct genetic phenomena as blood types and skin color.

Incomplete dominance

In some allele combinations, dominance does not exist. Instead, the two characteristics are equally expressed. For example, snapdragon flowers display **incomplete dominance** in their color. There are two alleles for flower color: one for white and one for red. When two alleles for white are present, the plant displays white flowers. When two alleles for red are present, the plant has red flowers. But when one allele for red is present with one allele for white, the color of the snapdragons is pink.

However, if two pink snapdragons are crossed, the phenotype ratio of the offspring is one red, two pink, and one white. These results show that the genes themselves remain independent; only the expressions of the genes appear to "blend." If the gene for red and the gene for white actually blended, pure red and pure white snapdragons could not appear in the offspring.

Multiple alleles

In certain cases, more than two alleles exist for a particular characteristic. Even though an individual has only two alleles, additional alleles may be present in the population. This condition is known as **multiple alleles.**

An example of multiple alleles occurs in blood type. In humans, blood groups are determined by a single gene with three possible alleles: A, B, or O. Red blood cells can contain two antigens, A and B. The presence or absence of these antigens results in four blood types: A, B, AB, and O. If a person's red blood cells have antigen A, the blood type is A. If a person's red blood cells have antigen B, the blood type is B. If the red blood cells have both antigen A and antigen B, the blood type is AB. If the red blood cells have neither antigen A nor antigen B, the blood type is O.

The alleles for type A and type B blood are examples of **codominance;** that is, both alleles are expressed. However, the allele for type O blood is recessive to both type A and type B. Because a person has only two of the three alleles, the blood type varies depending on which two alleles are present. For example, if a person has the A allele and the B allele, the blood type is AB. If a person has two A alleles, or one A and one O allele, the blood type is A. If a person has two B alleles, or one B and one O allele, the blood type is B. If a person has two O alleles, the blood type is O.

Polygenic inheritance

Although many characteristics are determined by alleles at a single place on the chromosome, some characteristics are determined by an interaction of genes on several chromosomes or at several places on one chromosome. This condition is **polygenic inheritance.**

An example of polygenic inheritance is human skin color. Genes for skin color are located in many places, and skin color is determined by which genes are present at these multiple locations. A person with many genes for dark skin will have very dark skin color, and a person with multiple genes for light skin will have very light skin color. Many people have some genes for light skin and some for dark skin, which explains why so many variations of skin color exist. Height is another characteristic probably reflecting polygenic characteristics.

Gene linkage

A chromosome has many thousands of genes; there are an estimated 20,000 genes in the human genome. Inheritance involves the transfer of chromosomes from parent to offspring through meiosis and sexual reproduction. It is common for a large number of genes to be inherited together if they are located on the same chromosome. Genes that are inherited together are said to form a linkage group. The concept of transfer of a linkage group is **gene linkage.**

Gene linkage can show how close two or more genes are to one another on a chromosome. The closer the genes are to each other, the higher the probability that they will be inherited together. Crossing over occurs during meiosis, but genes that are close to each other tend to remain together during crossing over.

Sex linkage

Among the 23 pairs of chromosomes in human cells, one pair is the **sex chromosomes.** (The remaining 22 pairs of chromosomes are referred to as **autosomes.**) The sex chromosomes determine the sex of humans. There are two types of sex chromosomes: the X chromosome and the Y chromosome. Females have two X chromosomes; males have one X and one Y chromosome. Typically, the female chromosome pattern is designated XX, while the male chromosome pattern is XY. Thus, the genotype of the human male would be 44 XY, while the genotype of the human female would be 44 XX (where 44 represents the autosomes).

In humans, the Y chromosome is much shorter than the X chromosome. Because of this shortened size, a number of sex-linked conditions occur. When a gene occurs on an X chromosome, the other gene of the pair probably occurs on the other X chromosome. Therefore, a female usually has two genes for a characteristic. In contrast, when a gene occurs on an X chromosome in a male, there is usually no other gene present on the short Y chromosome. Therefore, in the male, whatever gene is present on the X chromosome will be expressed.

An example of a sex-linked trait is colorblindness. The gene for colorblindness is found on the X chromosome. A woman is rarely colorblind because she usually has a dominant gene for normal vision on one of her X chromosomes. However, a male has the shortened Y chromosome; therefore, he has no gene to offset a gene for colorblindness on the X chromosome. As a result, the gene for colorblindness expresses itself in the male.

Another example of sex-linked inheritance is the blood disease hemophilia. In hemophilia, the blood does not clot normally because an important blood-clotting protein is missing. The gene for hemophilia occurs on the X chromosome. As females have two X chromosomes, one X chromosome usually has the gene for normal blood clotting. Therefore, the female may be a carrier of hemophilia but normally does not express hemophilia. Males have no offsetting gene on the Y chromosome, so the gene for hemophilia expresses itself in the male. This is why most cases of hemophilia occur in males.

Chapter Check-Out

Practice Project

1. Create a Punnett square and determine the genotypes and phenotypes of the resulting offspring if one parent is heterozygous for free earlobes (*Ee*) and the other parent is homozygous for attached earlobes (*ee*).

Questions

2. If both parents are heterozygous for the gene for free earlobes, what will be the ratio of phenotypes of the resulting offspring?
3. Two pink snapdragons were mated to each other. What are the resulting genotypic and phenotypic ratios of the offspring?

Answers

1. 2 *Ee* = free earlobes; 2 *ee* = attached earlobes.
2. The ratio will be 3:1, 3 free earlobes to 1 attached.
3. 1:2:1 genotypic ratio (*RR* = 1, *Rr* = 2, *rr* = 1); 1:2:1 phenotypic ratio (red = 1, pink = 2, white = 1).

Chapter 10

GENE EXPRESSION (MOLECULAR GENETICS)

Chapter Check-In

- ❑ Discovering the double helix
- ❑ Replicating DNA
- ❑ Synthesizing an RNA transcript
- ❑ Translating into an amino acid
- ❑ Understanding non-coding RNAs and chromatin modification

During the 1950s, a tremendous explosion of biological research occurred, and the methods of gene expression were elucidated. The knowledge generated during this period helped explain how genes function, and it gave rise to the science of molecular genetics. This science is based on the activity of deoxyribonucleic acid (DNA) and how this activity brings about the production of proteins in the cell. Genetic material is packaged into DNA molecules. DNA molecules relay the inherited information to messenger RNA (mRNA), which, in turn, codes for proteins. This chain of command is represented as:

$$\text{DNA} \rightarrow \text{mRNA} \rightarrow \text{protein}$$

The flow of information from DNA to protein is known as the central dogma of molecular biology.

In 1953, two biochemists, James D. Watson and Francis H. C. Crick, were working on a model for the structure of DNA. Once they saw the X-ray diffraction research done by the accomplished scientist Rosalind Franklin, the mystery of the structure of DNA was solved. Franklin's research helped elucidate that DNA consisted of two strands arranged in a double helix.

The publication of the structure of DNA opened a new realm of molecular genetics. Its structure provided valuable insight into how genes operate and how DNA can reproduce itself during mitosis, thereby passing on hereditary characteristics. Not only did the new research uncover many of the principles of protein synthesis, but it also gave rise to the science of biotechnology and genetic engineering (see Chapter 11).

DNA Structure

Deoxyribonucleic acid (DNA) consists of two long nucleotide chains. The two nucleotide chains twist around one another to form a double helix, a shape resembling a spiral staircase. Weak chemical bonds between the chains hold the two chains of nucleotides to one another.

A **nucleotide** in the DNA chain consists of three parts: a nitrogenous base, a phosphate group, and a molecule of deoxyribose. The **nitrogenous bases** of each nucleotide chain are of two major types: **purines** and **pyrimidines.** Purine bases have two fused rings of carbon and nitrogen atoms, while pyrimidine bases have only one ring. The two purine bases in DNA are adenine (A) and guanine (G). The pyrimidine bases in DNA are cytosine (C) and thymine (T).

The **phosphate group** of DNA is derived from a molecule of phosphoric acid. The phosphate group connects the deoxyribose molecules to one another in the nucleotide chain. **Deoxyribose** is a five-carbon carbohydrate. The purine and pyrimidine bases are attached to the deoxyribose molecules, and the purine and pyrimidine bases are opposite one another on the two nucleotide chains. Adenine is always opposite thymine and binds to thymine. Guanine is always opposite cytosine and binds to cytosine. Adenine and thymine are said to be complementary, as are guanine and cytosine. This is known as the principle of complementary base pairing.

DNA Replication

Before a cell enters the process of mitosis, its DNA replicates itself. Equal copies of the DNA pass into the daughter cells at the end of mitosis. In human cells, this means that 46 chromosomes (or molecules of DNA) replicate to form 92 chromosomes.

The process of **DNA replication** begins when specialized enzymes pull apart, or "unzip," the DNA double helix (see Figure 10-1). As the two strands separate, the purine and pyrimidine bases on each strand are exposed.

Figure 10-1 DNA replication. The double helix opens, and a complementary strand of DNA is synthesized along each strand.

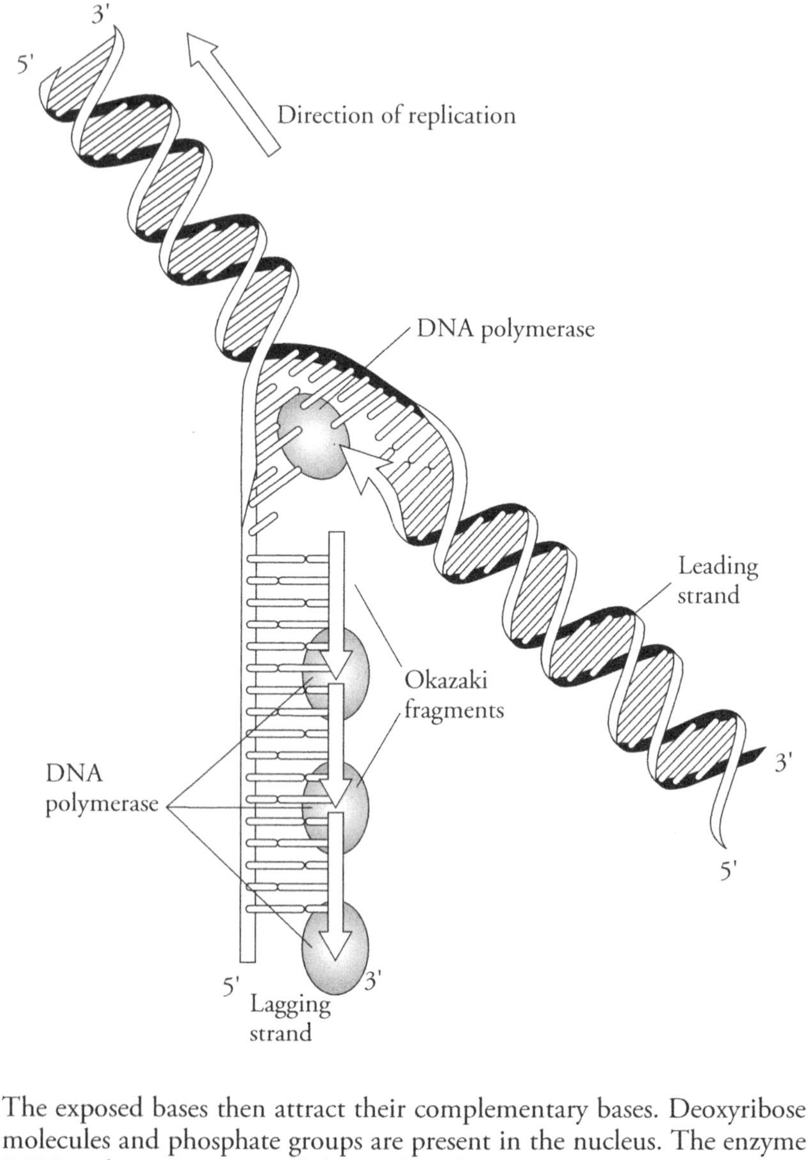

The exposed bases then attract their complementary bases. Deoxyribose molecules and phosphate groups are present in the nucleus. The enzyme **DNA polymerase** joins all the nucleotide components to one another, forming a long strand of nucleotides. Thus, the old strand of DNA directs the synthesis of a new strand of DNA through complementary base pairing.

The old strand then unites with the new strand to re-form a double helix. This process is called *semiconservative replication* because one of the old strands is conserved in the new DNA double helix.

DNA polymerase joins nucleotides in a 5'-3' direction on the leading strand, shown in Figure 10-1. However, DNA polymerase does not elongate a DNA strand in a 3'-5' direction. Therefore, the 3'-5' strand, called the lagging strand, is synthesized in short segments in a 5'-3' direction. These short segments placed on the lagging strand are **Okazaki fragments** and are ultimately joined together by the enzyme DNA ligase to form a new DNA strand.

DNA replication occurs during the S phase of the cell cycle. After replication has taken place, the chromosomal material shortens and thickens. The chromatids appear in the prophase of the next mitosis. The process then continues, and eventually two daughter cells form, each with the identical amount and kind of DNA as the parent cell. The process of DNA replication thus ensures that the molecular material passes to the daughter cells in equal amounts and types.

Though the process of DNA replication is amazingly accurate, mistakes do occur. These mutations are an important source of genetic variation, and viable mutations can be passed down to the offspring as heritable traits.

Protein Synthesis

During the 1950s and 1960s, it became apparent that DNA is essential in the synthesis of proteins. Among many other functions, **proteins** can serve as enzymes and as structural materials in cells. Many specialized proteins function in cellular activities. For example, in humans, the hormone insulin and the muscle cell filaments are composed of protein. The hair, skin, and nails of humans are composed of proteins, as are all the hundreds of thousands of enzymes in the body.

The structure of DNA determines the structure of proteins, and a protein's structure is essential to its proper function within a cell. For example, the protein hemoglobin is essential for oxygen transport around the circulatory system. It is composed of four separate protein chains, two α chains and two β chains. The amino acid sequence of each protein is specific in that it allows the overall protein to fold in a way necessary for its function. If there is a change in a single base (replacing a thymine with an adenine at a specific location) in the beta gene, the resulting amino acid is then valine instead of glutamic acid. Such a relatively small mistake leads to sickle cell anemia, a disease that causes a host of problems, all because of misfolded hemoglobin (brought on by a single wrong base!).

The key to a protein molecule is how the amino acids are linked. The sequence of amino acids in a protein is a type of code that specifies the protein and distinguishes one protein from another. A genetic code in the DNA determines this amino acid code. The genetic code consists of the sequence of nitrogenous bases in the DNA. How the nitrogenous base code is translated to an amino acid sequence in a protein is the basis for protein synthesis.

For protein synthesis to occur, several essential materials must be present, such as a supply of the 20 amino acids, which comprise most proteins. Another essential component is a series of enzymes that will function in the process. DNA and another form of nucleic acid called **ribonucleic acid (RNA)** are essential.

RNA is the nucleic acid that carries instructions from the nuclear DNA into the cytoplasm, where protein is synthesized. RNA is similar to DNA, with two exceptions. First, the carbohydrate in RNA is ribose rather than deoxyribose, and second, RNA nucleotides contain the pyrimidine uracil rather than thymine.

Types of RNA

In the synthesis of protein, three types of RNA function. The first type is called **ribosomal RNA (rRNA).** This form of RNA is used to manufacture ribosomes. **Ribosomes** are ultramicroscopic particles of rRNA and protein. They are the places (the chemical "workbenches") where amino acids are linked to one another to synthesize proteins. Ribosomes are found in large numbers along the membranes of the endoplasmic reticulum and in the cytoplasm of the cell (see Chapter 3).

A second important type of RNA is **transfer RNA (tRNA).** Transfer RNA exists in the cell cytoplasm and carries amino acids to the ribosomes for protein synthesis. When protein synthesis is taking place, enzymes link tRNA molecules to amino acids in a highly specific manner. For example, tRNA molecule X will link only to amino acid X; tRNA molecule Y will link only to amino acid Y.

The third form of RNA is **messenger RNA (mRNA).** In the nucleus, messenger RNA is constructed from DNA's code of base pairs and carries the code into the cytoplasm or to the rough endoplasmic reticulum, where protein synthesis takes place. Messenger RNA is synthesized in the nucleus using the DNA molecules. During the synthesis, the genetic information is transferred from the DNA molecule to the mRNA molecule. In this way, a genetic code can be used to synthesize a protein in a distant location. **RNA polymerase,** an enzyme, accomplishes mRNA, tRNA, and rRNA syntheses.

There are also non-coding RNA molecules (ncRNAs), which are not directly involved in protein synthesis. These will be further discussed in the section "Gene Control," later in this chapter.

Transcription

Transcription is one of the first processes in the mechanism of protein synthesis. In transcription, a complementary strand of mRNA is synthesized according to the nitrogenous base code of DNA. To begin, the enzyme RNA polymerase binds to an area of one of the DNA molecules in the double helix. (During transcription, only one DNA strand serves as a template for RNA synthesis. The other DNA strand remains dormant.) The enzyme moves along the DNA strand and "reads" the nucleotides one by one. Similar to the process of DNA replication, the new nucleic acid strand elongates in a 5'-3' direction, as shown in Figure 10-2. The enzyme selects complementary bases from available nucleotides and positions them in an mRNA molecule according to the principle of complementary base pairing. The chain of mRNA lengthens until a "stop" message is received.

Figure 10-2 The process of transcription. The DNA double helix opens, and the enzyme RNA polymerase synthesizes a molecule of mRNA according to the base sequence of the DNA template.

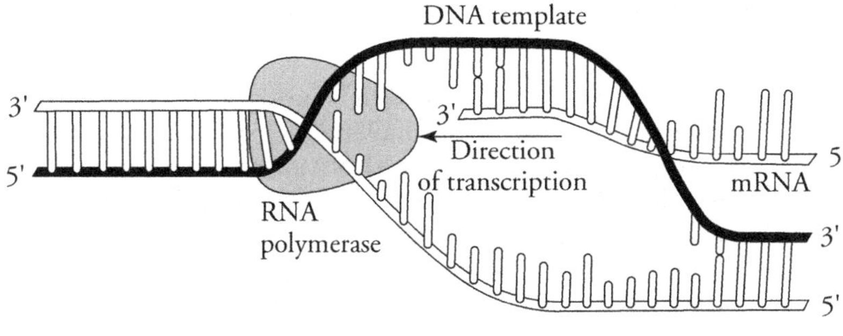

The nucleotides of the DNA strands are read in groups of three. Each group is a *codon.* Thus, a codon may be CGA, TTA, GCT, or any other combination of the four bases, depending on the codon's complementary sequence in the DNA strand. Each codon will later serve as a "code word" for an amino acid. First, however, the codons are transcribed to the mRNA molecule. Thus, the mRNA molecule consists of nothing more than a series of codons received from the genetic message in the DNA.

After the "stop" codon is reached, the synthesis of the mRNA comes to an end. The mRNA molecule leaves the DNA molecule, and the DNA molecule rewinds to form a double helix. Meanwhile, the mRNA molecule passes through a pore in the nucleus and proceeds into the cellular cytoplasm, where it moves toward the ribosomes located in the cytoplasm or on the rough endoplasmic reticulum.

Translation

The genetic code is transferred to an amino acid sequence in a protein through the **translation** process, which begins with the arrival of the mRNA molecule at the ribosome. While the mRNA is being synthesized, tRNA molecules are uniting with their specific amino acids according to the activity of specific enzymes. The tRNA molecules then begin transporting their amino acids to the ribosomes to meet the mRNA molecule.

After it arrives at the ribosomes, the mRNA molecule exposes its bases in sets of three, the codons. Each codon has a complementary codon called an **anticodon** on a tRNA molecule. When the codon of the mRNA molecule complements the anticodon on the tRNA molecule, the latter places the particular amino acid in that position. Then the next codon of the mRNA is exposed, and the complementary anticodon of a tRNA molecule matches with it. The amino acid carried by the second tRNA molecule is positioned next to the first amino acid, and the two are linked. At this point, the tRNA molecules release their amino acids and return to the cytoplasm to link up with new molecules of amino acid.

When it's time for the next amino acid to be positioned in the growing protein, a new codon on the mRNA molecule is exposed, and the complementary three-base anticodon of a tRNA molecule positions itself opposite the codon. This brings another amino acid into position, and that amino acid links to the previous amino acids. The ribosome moves farther down the mRNA molecule and exposes another codon, which attracts another tRNA molecule with its anticodon.

One by one, amino acids are added to the growing chain until the ribosome has moved down to the end of the mRNA molecule. Because of the specificity of tRNA molecules for their individual amino acids, and because of the base pairing between codons and anticodons, the sequence of codons on the mRNA molecule determines the sequence of amino acids in the protein being constructed. And because the codon sequence of the mRNA complements the codon sequence of the DNA, the DNA molecule ultimately directs the amino acid sequencing in proteins. The primary "start" codon on an mRNA molecule is AUG, which codes for the amino acid methionine. Therefore, each mRNA transcript begins with the AUG codon, and the resulting polypeptide begins with methionine.

Figure 10-3 shows that the process of protein synthesis begins with the production of mRNA (upper right). The mRNA molecule proceeds to the ribosome, where it meets tRNA molecules carrying amino acids (upper left). The tRNA molecule has a base code (anticodon) that complements the mRNA code (codon) and thereby brings a specific amino acid into position. The amino acids join together in peptide bonds (bottom), and the tRNA molecules are released to pick up additional amino acid molecules.

After the protein has been synthesized completely, it is removed from the ribosome for further processing and to perform its function. For example, the protein may be stored in the Golgi apparatus before being released by the cell, or it may be stored in the lysosome as a digestive enzyme. Also, a protein may be used in the cell as a structural component, or it may be released as a hormone, such as insulin. After synthesis, the mRNA molecule breaks up and the nucleotides return to the nucleus. The tRNA molecules return to the cytoplasm to unite with other molecules of amino acids, and the ribosome awaits the arrival of a new mRNA molecule.

Gene Control

The process of protein synthesis does not occur constantly in the cell. Rather, it occurs at intervals followed by periods of genetic "silence." Thus, the cell regulates and controls the gene expression process.

The control of gene expression may occur at several levels in the cell. For example, genes rarely operate during mitosis, when the DNA fibers shorten and thicken to form chromatin. The inactive chromatin is compacted and tightly coiled, and this coiling regulates access to the genes.

Other levels of gene control can occur during and after transcription. In transcription, certain segments of DNA can increase and accelerate the activity of nearby genes. After transcription has taken place, the mRNA molecule can be altered to regulate gene activity. For example, researchers have found that an mRNA molecule contains many useless bits of RNA that are removed in the production of the final mRNA molecule. These

Figure 10-3 The process of protein synthesis.

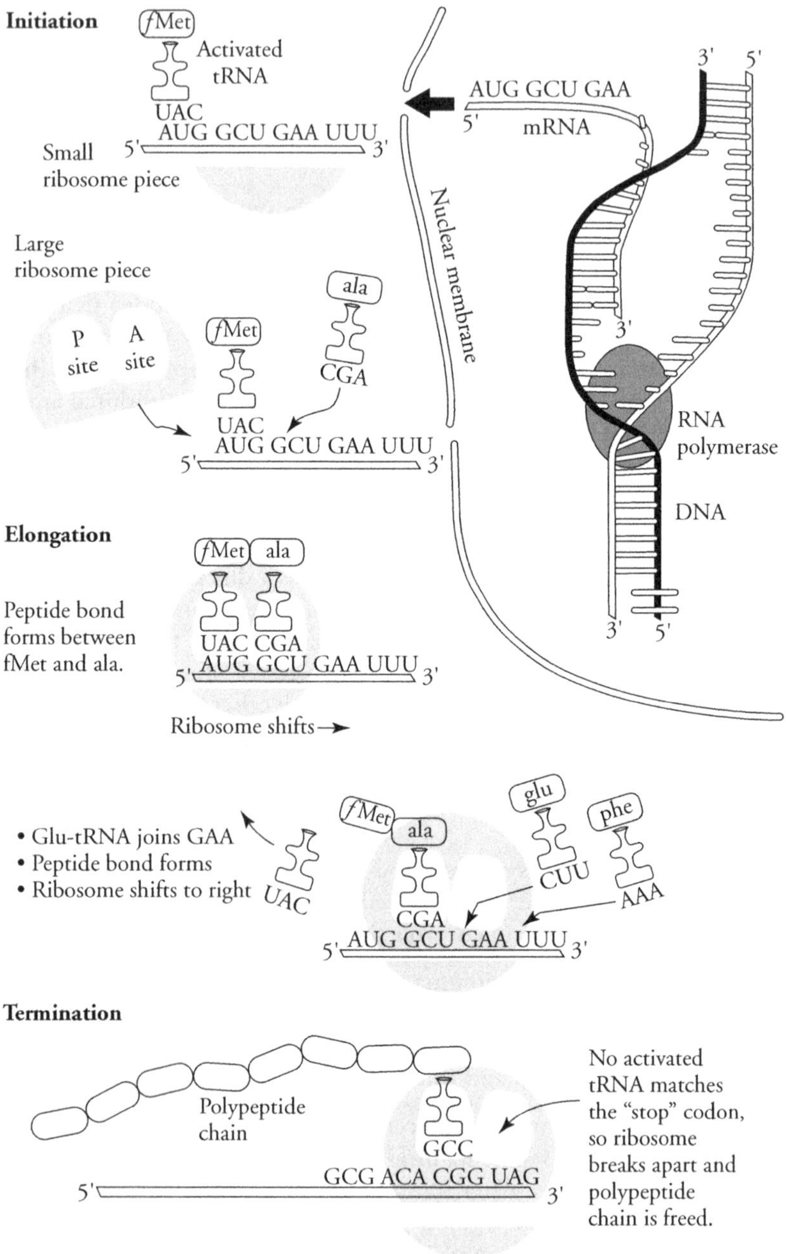

useless bits of nucleic acid are called *introns.* The remaining pieces of mRNA, called *exons,* are then spliced to form the final mRNA molecule. Thus, through removal of introns and the retention of exons, the cell can alter the message received from the DNA and control gene expression.

The concept of gene control has been researched thoroughly in bacteria. In these microorganisms, genes have been identified as structural genes, regulator genes, and control genes (or control regions). The three units form a functional unit called the *operon.*

The operon has been examined in close detail in certain bacteria. Scientists have found, for example, that certain carbohydrates can induce the presence of the enzymes needed to digest those carbohydrates. When lactose is present, bacteria synthesize the enzyme needed to break down the lactose. Lactose acts as the inducer molecule in the following way: In the absence of lactose, a regulator gene produces a repressor, and the repressor binds to a control region called the operator. This binding prevents the structural genes from encoding the enzyme for lactose digestion. When lactose is present, however, it binds to the repressor and thereby removes the repressor at the operator site. With the operator site free, the structural genes are free to produce their lactose-digesting enzyme.

The operon system in bacteria shows how gene expression can occur in relatively simple cells. The gene is inactive until it is needed and is active when it becomes necessary to produce an enzyme. Other methods of gene control are more complex and are currently being researched.

Non-coding RNAs (ncRNAs)

Protein-encoding mRNA is clearly an important molecule, but the other non-coding RNAs (ncRNAs) that control mRNA transcription are increasingly becoming the focus of attention. Two such ncRNA molecules are *microRNA* and *small interfering RNA.*

MicroRNAs (miRNAs) are small, single-stranded RNA molecules that, along with associated proteins, bind to complementary sequences in certain mRNA molecules. Once bound, these miRNAs block translation of the mRNA by either physically preventing the ribosome from binding or by causing the mRNA to degrade. The result—either blocked translation or mRNA degradation—depends on the extent of the base pairing between the miRNA and the target mRNA. Possibly half of human genes are regulated by miRNA.

Small interfering RNAs (siRNAs) are similar in structure and function to miRNAs. RNA interference (RNAi) is a means of disabling a gene by introducing siRNAs into a cell. Researchers employ RNAi to knock out specific genes in order to study their function.

Modification of chromatin structure

Recall that DNA is condensed and packaged with histone proteins into a complex known as chromatin (see Chapter 7). This compact structure helps DNA to fit into the nucleus and also provides opportunity for gene regulation. In order for a gene to be accessed by the transcriptional machinery, it must be "unwound" from the histone proteins. This is facilitated by certain enzymes adding acetyl groups ($-COCH_3$) to the histones **(histone acetylation).** Alternatively, if a segment of DNA needs to remain unexpressed (such as the inactivated mammalian X chromosomes), a different set of enzymes will add methyl groups ($-CH_3$) to certain bases, thus maintaining DNA's tightly wound and inaccessible form.

How a cell controls the expression of its genes is almost as important as the genes themselves. Every cell in a multicellular organism contains the same DNA (with the exception of the haploid gametes, which contain only a single set of chromosomes). Once cells begin to differentiate, they only express certain genes involved in their specific functions and processes. Differentiated cells are specialized at making tissue-specific proteins. A pancreatic cell, for example, will begin to express the gene for insulin (thus creating the protein needed for blood glucose regulation). Every other cell of the body also contains the gene for insulin production, yet the cells do not express this gene (e.g., it is not the function of cells in a hair follicle to make insulin). Many genes are, in fact, "turned off" via DNA modification. Modification of DNA and its associated histone proteins has a profound effect on a gene's expression. Furthermore, these modifications can be passed on to future generations and thus affect gene expression in progeny. This is called **epigenetic inheritance.** Alterations in normal modification have also been linked to some cancers, due to inappropriate gene expression.

Chapter Check-Out

Questions

1. What is the mRNA complement, from 5'-3', of the following DNA sequence?

   5'-GATCCGCTAAGGCCT-3'

2. What is the DNA complement to the DNA sequence above?
3. DNA replication takes place during ______ of the cell cycle.
   a. the G_1 phase
   b. metaphase
   c. the S phase
4. Which of the following components are NOT involved in translation?
   a. mRNA
   b. DNA
   c. ribosomes

Answers

1. 5'-AGGCCUUAGCGGAUC-3'
2. 5'-AGGCCTTAGCGGATC-3'
3. c
4. b

Chapter 11

RECOMBINANT DNA AND BIOTECHNOLOGY

Chapter Check-In

- ❑ Using biotechnology as a tool
- ❑ Detecting genetic disease
- ❑ Manufacturing proteins to fight disease
- ❑ Analyzing a genetic footprint

Biotechnology is a process that uses the scientific research on DNA for practical means. Biotechnology is synonymous with genetic engineering because the genes of an organism are changed during the process. Because the genes are changed, the DNA of the organism is said to be recombined. The result of the process is **recombinant DNA.** Recombinant DNA and biotechnology can be used to form proteins not normally produced in a cell to make drugs or vaccines or to promote human health. In addition, bacteria that carry recombinant DNA can be released into the environment under carefully controlled conditions to increase the fertility of the soil, serve as an insecticide, or relieve pollution. An organism that contains additional genes from another organism is said to be **transgenic.** Along with bacteria, transgenic plants and animals are also being created. Humans can also have the genes in their cells modified to produce proteins that relieve health-related deficiencies.

A researcher's ability to modify an organism's genome is possible because of genomics. **Genomics** is the study of an organism's entire set of chromosomes, including their function and species evolution. The sequencing of the human genome was completed in 2003, and since then, numerous prokaryotes (such as *E. coli*) and eukaryotes (both vertebrates and invertebrates) have been sequenced. This allows the comparison of genomes across a wide range of organisms, providing a better understanding of their genetic similarities and evolutionary history.

Tools of Biotechnology

The basic process of recombinant DNA technology involves manipulating an organism's DNA and thus altering the proteins being produced (see Chapter 10). During this synthesis, DNA provides the genetic code for the placement of amino acids in proteins. By intervening in this process, scientists can change the nature of the DNA, thereby changing the nature of the protein expressed by that DNA. By inserting genes into the genome of an organism, the scientist can induce the organism to produce a protein it does not normally produce.

The technology of recombinant DNA has been made possible in part by extensive research on microorganisms during the last half-century. One important microorganism in recombinant DNA research is *Escherichia coli,* commonly referred to as *E. coli.* The biochemistry and genetics of *E. coli* are well known, and its DNA has been isolated and made to accept new genes. The DNA can then be forced into fresh *E. coli* cells, and the bacteria will begin to produce the proteins specified by the foreign genes. Such altered bacteria are said to have been *transformed.*

Knowledge about viruses has also aided the development of DNA technology. **Viruses** are fragments of nucleic acid surrounded by a protein coat. Viruses attack cells and replicate within the cells, thereby destroying them. By attaching DNA to viruses, scientists use viruses to transport foreign DNA into cells and to connect it with the nucleic acid of the cells.

Another common method for inserting DNA into cells is to use **plasmids,** which are small loops of DNA in the cytoplasm of bacterial cells. Working with a plasmid is much easier than working with a chromosome, so plasmids are often the carriers, or **vectors,** of DNA. Plasmids can be isolated, recombined with foreign DNA, and then inserted into cells where they multiply as the cells multiply.

Interest in recombinant DNA and biotechnology heightened considerably during the 1960s and 1970s with the discovery of **restriction enzymes.** These enzymes catalyze the opening of a DNA molecule at a "restricted" point, regardless of the source of the DNA. Figure 11-1 shows that a human DNA molecule is opened at a certain site by the restriction enzyme EcoRI (upper left), and the desired DNA fragment is isolated (lower left). Plasmid DNA is treated with the same enzyme and opened. The DNA fragment is spliced into the plasmid to produce the recombinant DNA molecule.

Figure 11-1 Construction of a recombinant DNA molecule.

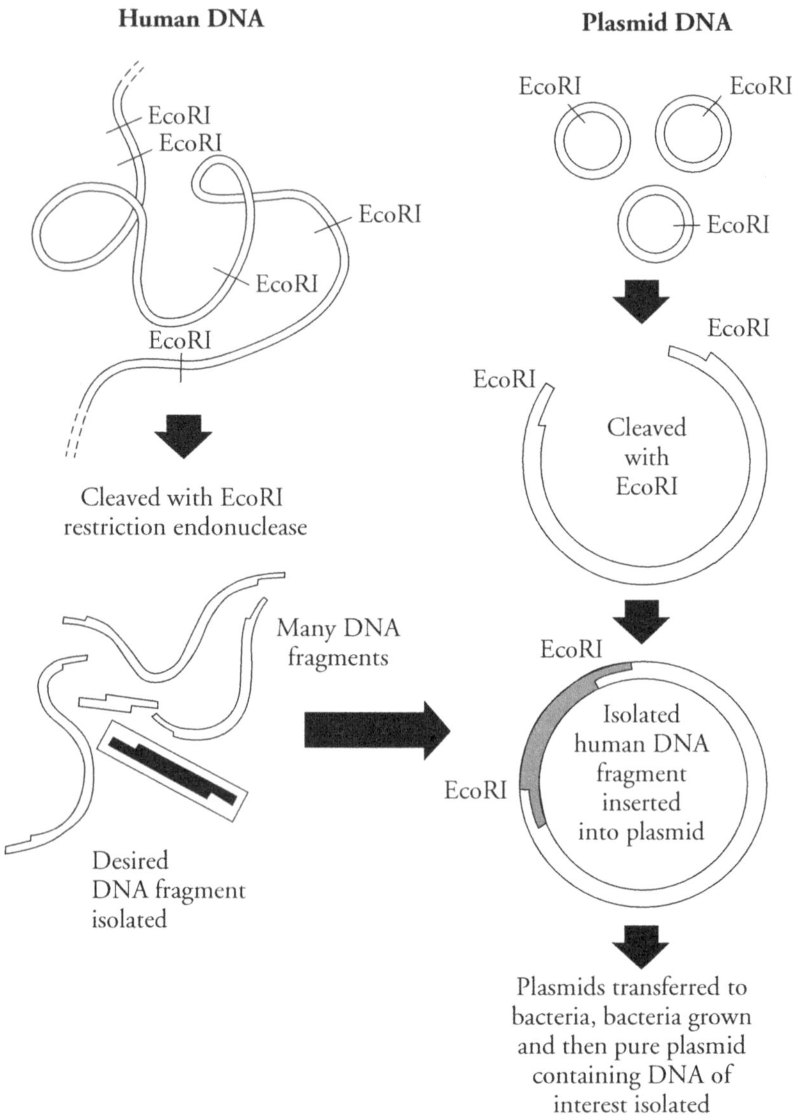

Certain restriction enzymes leave dangling ends of DNA molecules at the point where the DNA is opened. Foreign DNA can therefore be combined with the carrier DNA at this point. An enzyme called *DNA ligase*

forges a permanent link between the dangling ends of the DNA molecules at the point of union.

Recombinant DNA technology is sophisticated and expensive. Genes must be isolated, vectors must be identified, and gene control must be maintained. Stability of the vector within a host cell is important, and the scientist must be certain that nonpathogenic bacteria are used. Cells from mammals can be used to synthesize proteins, but cultivating these cells is difficult. In addition, the proper gene signals must be identified, RNA molecules must be bound to ribosomes, and the presence of introns must be considered. Collecting the gene product and exporting it from the cell are other considerations.

The genes used in DNA technology are commonly obtained from host cells or organisms called *gene libraries.* A gene library is a collection of cells identified as harboring a specific gene. For example, *E. coli* cells can be stored with the genes for human insulin in their chromosomes.

Pharmaceutical Products

Gene defects in humans can lead to deficiencies in proteins such as insulin, human growth hormone, and Factor VIII that may result in problems (diabetes, dwarfism, and impaired blood clotting, respectively). Proteins for these chemicals can now be replaced by proteins manufactured through biotechnology. For insulin production, two protein chains are encoded by separate genes in plasmids inserted into bacteria. The protein chains are then chemically joined to form the final insulin product. Human growth hormone is also produced within bacteria, but special techniques are used because the bacteria do not usually produce human proteins.

Therapeutic proteins, such as the following, can also be produced by biotechnology:

- *Tissue plasminogen activator (TPA),* a clot-dissolving protein, can now be produced in recombined mammalian cells.
- *Interferon,* an antiviral protein produced in *E. coli* cells, is currently used to fight certain types of cancers and for certain skin diseases.
- An *antisense molecule* is a molecule of RNA that reacts with and neutralizes the mRNA molecule used in protein synthesis. In doing so, the antisense molecule prevents the synthesis of a protein involved in a specific disease. For example, an antisense molecule can prohibit human host cells from producing key portions of the human immunodeficiency virus (HIV) when infection has occurred.

Vaccines represent another application of recombinant DNA technology. The hepatitis B vaccine now in use is composed of viral proteins manufactured by yeast cells and recombined with viral genes. The vaccine is safe because it contains no viral particles. Experimental vaccines against AIDS are being produced in the same way. One vaccine uses vaccinia (cowpox) virus as a vector. The virus has been combined with genes from several viruses, and it is hoped that injections of the vaccine will stimulate resistance to multiple diseases. Vaccines can also be produced by eliminating certain disease-inducing genes from a pathogen, leading to a harmless organism that will stimulate an immune response.

Diagnostic Testing

Recombinant DNA and biotechnology have opened a new era of diagnostic testing and have made detecting many genetic diseases possible. The basic tool of DNA analysis is a fragment of DNA called the *DNA probe.* A DNA probe is a relatively small, single-stranded fragment of DNA that recognizes and binds to a complementary section of DNA on a larger DNA molecule. After the probe unites with its target, it emits a signal like that from a radioactive isotope to indicate that a reaction has occurred.

To work effectively, a sufficiently large amount of target DNA must be available. To increase the amount of available DNA, a process called the **polymerase chain reaction (PCR)** is used. In a highly automated machine, the target DNA is combined with enzymes, nucleotides, and a primer DNA. In exponential fashion, the enzymes synthesize copies of the target DNA, so that in a few hours, billions of molecules of DNA exist where only a few were before.

Using DNA probes and PCR, scientists are now able to detect the DNA associated with HIV and AIDS. This has yielded a direct test for HIV that is preferred over the HIV antibody test. Lyme disease and genetic diseases such as cystic fibrosis, muscular dystrophy, Huntington's disease, and fragile X syndrome can be identified by DNA probes. (Cystic fibrosis is a respiratory disease in which mucus clogs the respiratory passageways and makes breathing difficult; muscular dystrophy is a disorder of the nervous system in which destruction of nerve fibers leads to erratic muscular activity; Huntington's disease is a disease of the nervous system accompanied by erratic movements and nerve degeneration; and fragile X syndrome is a disease of the X chromosome accompanied by a form of mental deficiency.)

Segments of DNA called **restriction fragment length polymorphisms (RFLPs)** are the objectives of the tests with gene probes. RFLPs are

apparently useless bits of DNA located near genes associated with the diseases. By locating RFLPs, the biotechnologist can locate the disease gene. DNA probes also detect microorganisms in the environment and identify viral and bacterial pathogens.

Gene Therapy

Gene therapy is a recombinant DNA process in which cells are taken from the patient, altered by adding genes, and replaced in the patient. The genes then provide the genetic codes for proteins the patient is lacking. Nonreproductive cells are used in gene therapy, so there is no carryover of inserted genes to the next generation.

In the early 1990s, gene therapy was used to correct a deficiency of the enzyme adenosine deaminase (ADA). Blood cells called lymphocytes were removed from the bone marrow of two children, then genes for ADA production were inserted into the cells using viruses as vectors. Finally, the cells were reinfused in the bodies of the children. Once established in the bodies, the gene-altered cells began synthesizing the enzyme ADA. Thus, the deficiency was removed and the disease resolved.

There are over 4,000 single-gene defects, and patients with these defects may be candidates for gene therapy. A multilayered review system now exists to ensure the safety of gene therapy proposals. Many aspects must be considered before approval is granted for gene therapy experiments.

Cloning and Stem Cells

Cloning refers to the ability to make a genetic replica of a cell (or even an entire organism, also called *reproductive cloning*). Interest in cloning is due primarily to its potential to create stem cells. A **stem cell** is a cell that has not yet differentiated and retains the ability to turn into many different tissues. Stem cells have enormous potential in medical applications because they can be used to replace diseased or damaged tissues and aid in organ repair or replacement.

In order for reproductive cloning to work, a differentiated cell must "dedifferentiate," allowing it to access the genes for any given cell type. This cell can then give rise to all the specialized cell types of an organism. One source of undifferentiated stem cells is from early-stage embryos. Considering the ethical implications of embryonic stem (ES) cells, research has begun to focus on adult stem cells. Adult stem cells are found in a number of tissues, including dental pulp, bone marrow, and the brain. These cells, however, are not able to differentiate into all cell types and thus are not as beneficial as embryonic stem cells.

In 2007, researchers announced their successful reprogramming of fully differentiated adult cells, thus inducing an ES-cell state. These induced pluripotent stem (iPS) cells can do everything an ES cell can do without the ethical implications associated with embryonic cells.

DNA Fingerprinting

The use of DNA probes and the development of retrieval techniques have made it possible to match DNA molecules to one another for identification purposes. This process has been used in a forensic procedure called **DNA fingerprinting.**

DNA fingerprinting depends on the presence of RFLPs, the repeating base sequences that exist in the human genome. As the pattern of RFLPs is unique for each individual, it can be used as a molecular fingerprint.

To perform DNA fingerprinting, DNA is obtained from an individual's blood cells, hair fibers, skin fragments, or other tissue. The DNA is then extracted from the cells and digested with enzymes. The resulting fragments are separated by a process called *electrophoresis,* in which electrical charges separate DNA fragments according to size. The separated DNA fragments are then detected with DNA probes and used to develop a fingerprint. A statistical evaluation enables the forensic pathologist to compare a suspect's DNA with the DNA recovered at a crime scene and to state with a high degree of certainty (usually 99 percent) that the suspect was at the crime scene.

Searching for DNA

The ability to retrieve DNA from ancient materials and museum specimens has given archaeologists and anthropologists hope of a glimpse at ancient life. Biochemists have successfully obtained DNA from extinct animals and plants. Evolutionary biologists have used the DNA to draw lineage patterns from the data. This often gives a better understanding of relationships between species. DNA isolated from ancient humans has been used to trace the movements of populations, such as the Anglo-Saxons, as well as to determine whether males were favored over females in certain societies.

Studies have also been performed on human origins by using the DNA found in the mitochondria. All of an offspring's mitochondrial DNA is derived from its mother. Because this DNA represents an unbroken line of genetic information, an analysis of mutation sites in the mitochondrial DNA can conceivably lead one back to the first human female.

DNA and Agriculture

Although plants are more difficult to work with than bacteria, gene insertions can be made into single plant cells. Then the cells can be cultivated to form a mature plant. The major method for inserting genes is through the plasmids of the bacterium called *Agrobacterium tumefaciens.* This bacterium invades plant cells, and its plasmids insert into plant chromosomes carrying the genes for tumor induction. Scientists remove the tumor-inducing genes and obtain a plasmid that unites with the plant cell without causing any harm.

Recombinant DNA and biotechnology have been used to increase the efficiency of plant growth by increasing the efficiency of the plant's ability to fix nitrogen. Scientists have obtained the genes for nitrogen fixation from bacteria and have incorporated those genes into plant cells. The plant cells can then perform a process that normally takes place only in bacteria.

DNA technology has also been used to increase plant resistance to disease by reengineering the plant to produce viral proteins. Also, the genes for an insecticide obtained from a bacterium have been inserted into plants to allow the plants to resist caterpillars and other pests.

One of the first agricultural products of biotechnology was the rot-resistant tomato. This plant was altered by adding a gene that produces an antisense molecule. The antisense molecule inhibits the tomato from producing the enzyme that encourages rotting. Without this enzyme, the tomato can ripen longer on the vine.

Transgenic Animals

A **transgenic animal** is an animal in which one or more genes have been introduced into its nonreproductive cells. The first transgenic animal was produced in 1983 when genes for human growth hormone were introduced into mice.

Transgenic animals can be used to produce valuable products. For example, a transgenic pig has been produced with the ability to synthesize human hemoglobin for use as a blood substitute. Also, a transgenic cow has been bred with the ability to produce human lactoferrin, an iron-building milk protein and a potential antibacterial agent. A transgenic sheep can synthesize a protein that helps emphysema patients breathe more easily, and a transgenic goat has been developed to produce a protein needed by cystic fibrosis patients.

Human Genome

In 1990, researchers at Celera Genomics and at the National Human Genome Research Institute began an ambitious endeavor to sequence the entire human genome. In 2003, the project was completed, resulting in the sequencing of all human chromosomes. The Human Genome Project revealed that the human genome contains only 20,000 to 25,000 genes. This estimation, based on the sequence data, is substantially below previous predictions. The sequence data has led to the estimation that only about 1 percent of the human genome actually encodes functional proteins. Once the jigsaw puzzle is completed, the data will undoubtedly help researchers devise new diagnostics and treatments for genetic diseases.

In addition to sequencing the human genome, researchers have sequenced the genomes of *Drosophila melanogaster* (fruit fly), *Arabidopsis thaliana* (plant), *Saccharomyces cerevisiae* (budding yeast), and *C. elegans* (worm). In addition, mouse, rat, and zebrafish genomes have been sequenced. Eukaryotic organisms are also useful to the research community. The genome of *Plasmodium* (the organism that causes malaria) has also been sequenced. The goals of these sequencing projects are to prepare gene linkage maps and physical maps. A **gene linkage map** pinpoints the location of genes based on their connection to certain marker gene sequences. A **physical map,** in comparison, gives the actual number of bases between genes on a chromosome; therefore, it locates the gene of interest more precisely.

Ultimately, scientists hope to learn the actual names and sequences of all 3 billion nitrogenous base pairs in the human genome. Automation and computerization are essential tools in the sequencing, and the development of the specific technology is underway.

Chapter Check-Out

Questions

1. An enzyme called ______ links the ends of DNA molecules, forming a permanent bond.
   - **a.** restriction
   - **b.** polymerase
   - **c.** ligase
   - **d.** virus
2. A technique used to amplify a specific fragment of DNA is called ______.
   - **a.** RFLP
   - **b.** transgenic
   - **c.** ligation
   - **d.** PCR
3. DNA was extracted from an individual's skin cells. The DNA was digested using restriction enzymes and resolved by electrophoresis. The fragments were analyzed to develop a pattern. This technique is called ______.
   - **a.** PCR
   - **b.** RFLP
   - **c.** gene therapy
   - **d.** DNA probe

Answers

1. c
2. d
3. b

Chapter 12
PRINCIPLES OF EVOLUTION

Chapter Check-In

- ❑ Evolving by natural selection
- ❑ Validating the theory of evolution
- ❑ Driving the evolutionary process

Evolution implies a change in one or more characteristics in a population of organisms over a period of time. The concept of evolution is as ancient as Greek writings, where philosophers speculated that all living things are related to one another, although remotely. The Greek philosopher Aristotle perceived a "ladder of life," where simple organisms gradually change to more elaborate forms. Opponents of this concept were led by several theologians who pointed to the biblical account of creation as set forth in the Book of Genesis. One prelate, James Ussher, calculated that creation had taken place on October 26, 4004 B.C.E., at 9 a.m.

Opponents of the creationist argument were encouraged by geologists who postulated that Earth is far older than 4,004 years. In 1785, James Hutton postulated that Earth was formed by an ancient progression of natural events, including erosion, disruption, and uplift. In the early 1800s, Georges Cuvier suggested that Earth was 6,000 years old, based on his calculations. In 1830, Charles Lyell published evidence pushing the age of Earth back several million years.

Amid the controversy over geology and Earth's age, French zoologist Jean-Baptiste Lamarck suggested a theory for evolution based on the development of new traits in response to a changing environment. For example, the neck of the giraffe stretched as it reached for food. Lamarck's theory of "use and disuse" gained favor, and his concept of "acquired characteristics" was accepted until the time of Charles Darwin, many years later.

Charles Darwin was the son of an English physician. As a naturalist on the ship H.M.S. *Beagle,* Darwin traveled to remote regions of South America and other destinations. His observations on this trip led him to develop his own theory of evolution. Darwin was particularly interested in the finches and tortoises of the Galapagos Islands. He pondered how different species of animals could have developed on this remote set of islands 200 miles west of Ecuador.

Darwin returned to England from South America in 1838 and continued to ponder the theory of evolution. He was influenced by Thomas Malthus' *An Essay on the Principle of Population.* In his book, Malthus pointed out the human population's continual struggle for survival and that a population's natural tendency is to produce more offspring than can possibly survive. Darwin applied this principle to animals and plants, and his theory of evolution began to develop.

In 1858, another English naturalist, Alfred Russel Wallace, developed a concept of evolution similar to Darwin's. Wallace wrote a paper on the subject and corresponded with Darwin. The two men decided to simultaneously present papers on evolution to London's scientific community in 1858. The next year, 1859, Darwin published his famous book, *On the Origin of Species by Means of Natural Selection, or the Preservation of Favoured Races in the Struggle for Life.* The book has become known simply as *The Origin of Species.*

Theory of Evolution

In his book *The Origin of Species,* Darwin presented evidence for his "descent with modification" theory, which has come down to us as the theory of evolution, although Darwin avoided the term "evolution." Essentially, Darwin suggested that random variations take place in living things and that the environment selects those individuals better able to survive and reproduce. The method of selecting individuals is known as **natural selection.** The selected individuals are more likely to pass on their traits to their offspring, and the population continues to evolve.

Two essential points underlie natural selection. First, the genetic variations that take place in living things are random variations. Second, the genetic variations are small and cause little effect relative to a given population. Over time, these small genetic variations lead to the gradual development of a species rather than the sudden development of a species. Darwin proposed that variations appear without direction and without design. He assumed that among inherited traits, some traits were better than others. If an inherited trait provided an advantage over another, it would provide a reproductive advantage to the bearer of the trait. Thus, if long-necked giraffes could reach food better than short-necked giraffes, the long-necked giraffes would survive, reproduce, and yield a population consisting of more long-necked giraffes.

As the central concept of Darwin's theory of evolution, natural selection implies that the fittest survive and spread their traits through a population. This concept is called the **survival of the fittest.** The fitness implied is reproductive fitness; that is, the ability to survive in the environment and propagate the species. Natural selection serves as a sieve to remove the unfit from a population and allow the fittest to reproduce and continue the population. Today, scientists know that other factors also influence evolution. Evolution and common ancestry are supported by multiple lines of evidence, such as similarities in DNA sequences, anatomical structures, and embryonic development.

Evidence for Evolution

In his book, Darwin offered several pieces of evidence that supported evolution. He attempted to convince the scientific community of the validity of his theory.

Paleontology

One piece of evidence offered by Darwin is found in the science of paleontology. **Paleontology** deals with locating, cataloging, and interpreting the life forms that existed in past millennia. It is the study of fossils—the bones, shells, teeth, and other remains of organisms, or evidence of ancient organisms, that have survived over eons.

Paleontology supports the theory of evolution because it shows a descent of modern organisms from common ancestors. Paleontology indicates that fewer kinds of organisms existed in past eras, and the organisms were probably less complex. As paleontologists descend deeper and deeper into layers of rock, the variety and complexity of fossils decrease. The fossils from the uppermost rock layers are most like current forms. Fossils from the deeper layers are the ancestors of modern forms.

Comparative anatomy

More evidence for evolution is offered by **comparative anatomy** (see Figure 12-1). As Darwin pointed out, the forelimbs of such animals as humans, porpoises, bats, and other creatures are strikingly similar, even though the forelimbs are used for different purposes (that is, lifting, swimming, and flying, respectively). Darwin proposed that similar forelimbs have similar origins, and he used this evidence to point to a common ancestor for modern forms. He suggested that various modifications are nothing more than adaptations to the special needs of modern organisms.

Figure 12-1 The forelimbs of a human and four animals showing the similarity in construction. This similarity was offered by Darwin as evidence that evolution has occurred.

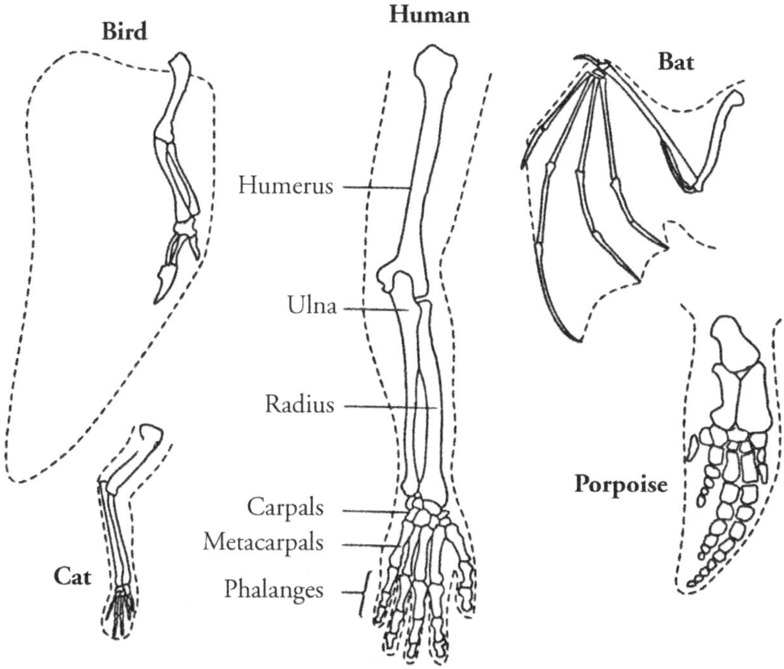

Darwin also observed that animals have structures they do not use. Often these structures degenerate and become undersized compared with similar organs in other organisms. The useless organs or body parts are called *vestigial organs.* In humans, they include the appendix, the fused tail vertebrae, the wisdom teeth, and muscles that move the ears and nose. Darwin maintained that vestigial organs may represent structures that have not quite disappeared. Perhaps an environmental change made the organ unnecessary for survival, and the organ gradually became nonfunctional and reduced in size. For example, the appendix in human ancestors may have been an organ for digesting certain foods, and the coccyx at the tip of the vertebral column may be the remnants of a tail possessed by an ancient ancestor.

Embryology

Darwin noted the striking similarity among embryos of complex animals such as humans, chickens, frogs, reptiles, and fish. He wrote that the uniformity is evidence for evolution. He pointed out that human embryos pass through a number of embryonic stages inherited from their ancestors because they have inherited the developmental mechanisms from a common ancestor. These mechanisms are modified in a way that is unique to an organism's way of life.

The similarities in comparative embryology are also evident in the early stages of development. For example, fish, bird, rabbit, and human embryos are similar in appearance in the early stages. They all have gill slits, a two-chambered heart, and a tail with muscles to move it. Later on, as the embryos grow and develop, they become less and less similar. The branch of biology that focuses on embryos and their development is called **embryology.**

Comparative biochemistry

Although the biochemistry of organisms was not well known in Darwin's time, modern biochemistry indicates there is a biochemical similarity in all living things. This comparison of biochemical processes with ancient species is called **comparative biochemistry.** For example, the same mechanisms for trapping and transforming energy and for building proteins from amino acids are nearly identical in almost all living systems. DNA and RNA are the mechanisms for inheritance and gene activity in all living organisms. The structure of the genetic code is almost identical in all living things. This uniformity in biochemical organization underlies the diversity of living things and points to evolutionary relationships.

Domestic breeding

From observing the domestic breeding experiments of animal and plant scientists, Darwin developed an idea about how evolution takes place. **Domestic breeding** brings about new forms that differ from ancestral stock. For example, pigeon fanciers have developed many varieties of pigeons through domestic breeding experiments. In effect, evolution has taken place under the guidance of human hands. The development of new agricultural crops by farmers and botanists provides more evidence for directed evolution.

Geographic distribution

Darwin was particularly interested in the life forms of the Galapagos Islands. He noticed how many of the birds and other animals on the islands were found only there. The finches were particularly puzzling; Darwin found 13 species of finches not found anywhere else in the world, as far as he knew. He concluded that the finches had evolved from a common ancestral group that probably reached the islands many generations earlier. In the isolation of the Galapagos Islands, the original finches had probably evolved into the 13 species.

The **geographic distribution** of species in geographic areas can help to explain evolution. For example, alligators are located only in certain regions of the world, presumably because they have evolved in those regions. The islands of Australia and New Zealand have populations of animals found nowhere else in the world because of their isolated environments.

Mechanisms of Evolution

Populations evolve, but individual organisms do not. A **population** is an interbreeding group of individuals of one species in a given geographic area at the same time. A population evolves because the population contains the collection of genes called the **gene pool.** As changes in the gene pool occur, a population evolves.

Mutation

Mutation, a driving force of evolution, is a random change in an organism's genetic makeup, which influences the population's gene pool. It is a change in the nature of the DNA in one or more chromosomes. Mutations give rise to new alleles; therefore, they are a source of genetic variation in a population.

Mutations may be harmful or benign, but they may also be beneficial. For example, a mutation may permit organisms in a population to produce enzymes that will allow them to use certain food materials. Over time, these types of individuals survive, while those that don't have the mutations are more likely to perish. Therefore, natural selection tends to remove the less-fit individuals, allowing more-fit individuals to survive and form a population.

Gene flow

Another mechanism of evolution may occur during the migration of individuals from one group or location to another. When the migrating individuals interbreed with the new population, they contribute their genes to the gene pool of the local population. This establishes **gene flow** in the population.

Gene flow occurs, for example, when wind carries seeds far beyond the bounds of the parent plant population. As another example, animals may be driven off from a herd. This forces them to migrate to a new population, thereby bringing new genes to a gene pool. Gene flow tends to increase the similarity between remaining populations of the same species because it makes gene pools more similar to one another.

Genetic drift

Another mechanism for evolution is **genetic drift,** which can occur when a small group of individuals leaves a population and establishes a new one in a geographically isolated region. For example, when a small population of fish is placed in a lake, the fish population will evolve into one that is different from the original. Fitness of a population is not considered in genetic drift, nor does genetic drift occur in a very large population.

Natural selection

Another mechanism for evolution is natural selection, which occurs when populations of organisms are subjected to the environment. The fittest creatures are more likely to survive and pass their genes to their offspring, producing a population that is better adapted to the environment. The genes of less-fit individuals are less likely to be passed on to the next generation. The important selective force in natural selection is the environment.

Environmental fitness may be expressed in several ways. For example, it may involve an individual's ability to avoid predators, it may imply a greater resistance to disease, it may enhance the ability to obtain food, or it may mean resistance to drought. Fitness may also be measured as enhanced reproductive ability, such as the ability to attract a mate. Better-adapted individuals produce relatively more offspring and pass on their genes more successfully than less-adapted individuals.

Several types of natural selection appear to affect populations. One type, *stabilizing selection,* occurs when the environment selects against organisms of a population with extreme versions of a trait. Another type of natural selection is *disruptive selection.* Here, the environment favors extreme types in a population at the expense of intermediate forms, thereby splitting the population into two or more subpopulations. A third type of natural selection is *directional selection.* In this case, the environment selects for an extreme characteristic. The development of antibiotic-resistant bacteria in the modern era is an example of directional selection.

Species development

A **species** is a group of individuals that share a number of features and are able to interbreed with one another, producing fertile (non-sterile) offspring. (When individuals of one species mate with individuals of a different species, any offspring are usually sterile.) A species is also defined as a population whose members share a common gene pool.

The evolution of a species is *speciation,* which can occur when a population is isolated by geographic barriers, such as occurred in the isolation of Australia, New Zealand, and the Galapagos Islands. The variety of life forms found in Australia but nowhere else is an example of speciation by geographic barriers.

Speciation can also occur when reproductive barriers develop. For example, when members of a population develop anatomical barriers that make mating with other members of the population difficult, a new species can develop. The timing of sexual activity is another example of a reproductive barrier. Spatial difference, such as one species inhabiting treetops while another species lives at ground level, is another reason why species develop.

Gradual versus rapid change

Darwin's theory included the observation that evolutionary changes take place slowly. In many cases, the fossil record shows that a species changed gradually over time. The theory that evolution occurs gradually is known as *gradualism.*

In contrast to gradualism is the theory of *punctuated equilibrium,* which is a point of discussion among scientists. According to the theory of punctuated equilibrium, some species have long, stable periods of existence interrupted by relatively brief periods of rapid change.

Both groups of scientists agree that natural selection is the single most important factor in evolutionary changes in species. Whether the change is slow and gradual or punctuated and rapid, one thing is certain: Organisms have evolved over time.

Shifts in allelic frequency

The theory of natural selection states that organisms with an advantageous (and heritable) trait will increase in number compared to their less-fit counterparts. By examining the allele frequencies in a population, a change in those percentages over time shows evolution, with advantageous heritable traits increasing in proportion. Since the environment dictates which alleles are considered "most fit," it makes sense that a change in the environment would lead to a change in the allelic composition (the relative percentages of alleles). For example, suppose in 30 years (from 1980 to 2010), a species of tree with dark bark increased in number in a particular forest. There is a corresponding change in the gene pool squirrels that also live in this forest. Figure 12-2 shows the shift in the allele frequencies of squirrel fur color between 1980 and 2010.

Figure 12-2 The relative percentages of light fur and dark fur alleles in a population of squirrels, before (1980) and after (2010) an increase in dark-colored trees in the squirrels' habitat.

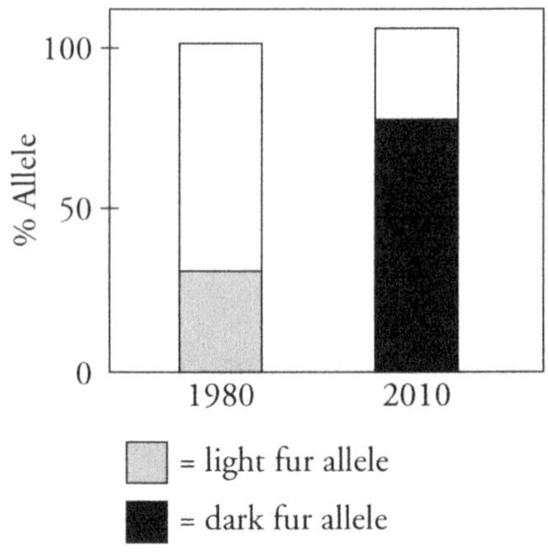

The dark allele has become the advantageous trait (dark-furred squirrels are better camouflaged against dark trees), and thus the percentage of the dark allele has increased. This shift in allele percentages shows a quantifiable result of evolution.

Chapter Check-Out

Questions

1. All of the following techniques have been commonly used to validate Darwin's theory of evolution EXCEPT ______.
   a. RFLP
   b. paleontology
   c. comparative anatomy
2. Wind carrying seeds to a new, uninhabited site is an example of ______.
   a. gene flow
   b. genetic drift
   c. species development
3. A small group that leaves a population and establishes a new one in a geographically isolated region is an example of ______.
   a. gene flow
   b. genetic drift
   c. speciation

Answers

1. a
2. a
3. b

Chapter 13
THE ORIGIN AND EVOLUTION OF LIFE

Chapter Check-In

- ❑ Forming organic molecules
- ❑ Tracing the first eukaryotic cells
- ❑ Exploring the Paleozoic, Mesozoic, and Cenozoic eras

Scientists hypothesize that the universe came into existence about 15 billion years ago with a colossal explosion often referred to as the *big bang*. The gases and dust from that explosion produced the earliest generation of stars, and over a period of billions of years, the stars exploded, and their debris formed other stars and planets. The solar system was presumably formed in this way 4 to 5 billion years ago. During the next billion years, the molten earth cooled, forming a hardened, outer crust. About 3.5 billion years ago, living things came into being.

Origin of Organic Molecules

About 3.8 billion years ago, Earth's atmosphere consisted of such elements as nitrogen, hydrogen, sodium, sulfur, and carbon. Some of these elements combined to form hydrogen sulfide, methane, water, and ammonia. Water vapor condensed, resulting in millions of years of torrential rains, during which the oceans formed. Gas from Earth's core came to the surface through volcanoes. Ultraviolet radiation bathed Earth, and the elements and compounds interacted with one another to form complex molecules.

In 1953, Stanley Miller and Howard Urey performed a classic experiment in which they circulated methane, ammonia, water vapor, and hydrogen gas in a closed environment and passed electric sparks through it. After several days, they discovered that complex compounds of carbon had formed in the mixture. Their experiments indicated that in the primitive

Earth atmosphere, complex organic molecules could form, including amino acids. The hypothesis they expressed is known as *primordial soup.*

Recent hypotheses about the origin of organic molecules suggest that these molecules may have formed in hydrothermal vents deep in the oceans, where hot gases and elements emerge from cracks in Earth's crust. Living organisms have been found near these vents, lending credence to the hypothesis.

Origin of Cells

The appearance of the first cells marked the origin of life on Earth. However, before cells could form, the organic molecules must have united with one another to form more complex molecules called *polymers.* Examples of polymers are polysaccharides and proteins.

In the 1950s, Sidney Fox placed amino acids in primitive Earth conditions and showed that amino acids would unite to form polymers called **proteinoids.** The proteinoids were apparently able to act as enzymes and catalyze organic reactions.

More recent evidence indicates that RNA molecules have the ability to direct the synthesis of new RNA molecules as well as DNA molecules. Because DNA provides the genetic code for protein synthesis, it is conceivable that DNA may have formed in the primitive Earth environment as a consequence of RNA activity. Then DNA activity could have led to protein synthesis (see Chapter 10).

For a cell to come into being, some sort of enclosing membrane is required to hold together the organic materials of the cytoplasm. A generation ago, scientists believed that membranous droplets formed spontaneously. These membranous droplets, called **protocells,** were presumed to be the first cells. Modern scientists believe, however, that protocells do not carry any genetic information and lack the internal organization of cells. Thus the protocell perspective is not widely accepted. Several groups of scientists are currently investigating the synthesis of polypeptides and short nucleic acids on the surface of clay. The origin of the first cells remains a mystery.

Ancient Life

Earth came into existence about 4.6 billion years ago, and about 3.8 billion years ago, the chemical composition of Earth's surface began to change. Scientists estimate that at about 3.5 billion years ago, the first cells were in existence.

Scientists believe that the first cells lived within the organic environment of Earth and used organic foods for their energy. The type of chemistry in those first cells was somewhat similar to **fermentation,** which uses organic molecules such as glucose. The energy yield, although minimal, is enough to sustain simple living things (see Chapter 6). However, organic material would soon have been used up if this were the sole source of nutrition, so a new process had to develop.

The evolution of a pigment system that could capture energy from sunlight and store it in chemical bonds was an essential evolutionary breakthrough. The organisms that possess these pigments are commonly referred to as **cyanobacteria** (at one time, they were called blue-green algae). These single-celled organisms produce carbohydrates by the process of photosynthesis. In doing so, they produce oxygen as a waste product (see Chapter 5). For about 1 billion years, photosynthesis provided oxygen to the atmosphere, which gradually changed until it became oxygen-rich, as it is today.

Another group of organisms that was present at the same time as the cyanobacteria was a group of bacteria called **archae.** Archae differ from other species of bacteria (known as kingdom **eubacteria** and representative of domain Bacteria) in that archae have a different ribosomal structure, different cell membrane composition, and different cell wall composition. The archae have been traced to a period of about 3 billion years ago. They are able to multiply at the very high temperatures that were present on Earth then, and their nutritional requirements reflect the composition of primitive Earth.

First Eukaryotes

The cyanobacteria and archae of primitive Earth are also referred to as **prokaryotes** (together with the eubacteria). Prokaryotes are discussed in Chapter 16. Approximately 1.5 billion years ago, in an oxygen-containing atmosphere, the first **eukaryotes** came into being. Eukaryotes have a nucleus, a nuclear membrane, a number of organelles, a ribosomal structure different from that of prokaryotes, cell division by mitosis, and other features that distinguish them from prokaryotes (see Chapter 3).

No one is certain how eukaryotes came into being. The *endosymbiotic theory* suggests that bacteria were engulfed by larger cells. The bacterial cells remained in the cell, assumed some of the chemical reactions for these cells, and became the mitochondria of these cells. The cells then reproduced and flourished, becoming animal cells.

An extension of the endosymbiotic theory refers to plants. In this case, pigmented bacteria, such as the cyanobacteria, were engulfed by larger cells. The cyanobacteria remained in the cells and became the chloroplasts of these cells. Photosynthesis occurs in the chloroplasts of plant cells and some protists.

Life on Land

For billions of years, the only life present on Earth existed in the nutrient-rich environments of the oceans, lakes, and rivers. About 600 million years ago, the Paleozoic Era began. Scientists believe that living things first came to occupy land during this era. They also believe that during a subdivision of the Paleozoic Era called the Cambrian Period, the main groups of marine invertebrates in existence today evolved. A so-called "Cambrian explosion" occurred. The appearance of multicellular organisms is notable in the Cambrian Period, when evolution by natural selection led to a vast array of organisms filling every conceivable niche on Earth. Many organisms that arose at that time have since become extinct.

After the Cambrian Period came the Ordovician Period. In the Ordovician Period, wormlike animals with stiff rods along their backs came into being. These organisms, now called **chordates,** include reptiles, amphibians, birds, and mammals. A backbone composed of vertebrae was first developed during the next period, the Silurian Period. During the Devonian Period that followed, bony fish developed. The first terrestrial plants also evolved at about this time.

The next most recent era after the Paleozoic Era is the Mesozoic Era, which began about 250 million years ago. During this era, reptiles such as the dinosaurs evolved and became the predominant life form on Earth. A mammal-like animal also evolved during this period. After the dinosaurs died out at the end of the Mesozoic Era, the mammal-like animal evolved into the modern **mammals.** Birds first appeared during the Mesozoic Era, and lush greenery spread over Earth.

During the Mesozoic Era, the continents existed as one landmass called Pangaea. Toward the end of the era, the landmass broke into smaller pieces, and the pieces moved apart to form the continents. The movement of the continents isolated plants and animals from one another, sparking the **genetic drift** that gives us different species of animals and plants in different parts of the world today.

At the end of the Mesozoic Era, a great extinction took place, and the dinosaurs disappeared. Although the reasons for the extinction are not clear, many scientists believe that a great meteor collided with Earth, presumably in the Yucatan Peninsula area of Mexico. The dust raised by this impact so darkened Earth that photosynthesis became impossible.

The amount of available food became scarce, and the dinosaurs disappeared. Mammals filled the void left by the dinosaurs' absence.

The Cenozoic Era, the most recent era, has been called the age of mammals. The era began about 65 million years ago and extends to the present time. Mammals that survived the great extinctions at the end of the Mesozoic Era radiated throughout Earth and filled the niches once inhabited by the dinosaurs. Primates first appeared on Earth about 35 million years ago, and modern humans are believed to have evolved about 40,000 years ago.

Chapter Check-Out

Questions

1. Sidney Fox showed that amino acids could unite to form polymers called ______ that act as enzymes.
   a. protocells
   b. eukaryotes
   c. proteinoids
2. Organisms called ______ evolved a pigment system that could capture energy from sunlight to create energy.
   a. eukaryotes
   b. cyanobacteria
   c. archae
3. According to the endosymbiotic theory, bacterial cells ultimately became ______ within eukaryotic cells.
   a. nuclei
   b. mitochondria
   c. Golgi bodies
4. The ______ Era is known as the age of the mammals, whereas the ______ Era is when reptiles, such as the dinosaurs, evolved to become the predominant life form on Earth.
   a. Cenozoic; Paleozoic
   b. Paleozoic; Cenozoic
   c. Cenozoic; Mesozoic

Answers

1. c
2. b
3. b
4. c

Chapter 14
HUMAN EVOLUTION

Chapter Check-In

- ❑ Walking upright
- ❑ Using tools
- ❑ Developing language
- ❑ Building societies

Fossils and fragments of jaws suggest to scientists that the ancestors of monkeys, apes, and humans began their evolution approximately 50 million years ago. Additional evidence for this evolution has been derived from experiments in biochemistry and changes that occur in the DNA of cells. Mutation rates in the DNA have been calculated, and evolutionary changes can be determined from these calculations. By noting mutation rates, scientists can estimate the time since two species diverged from a single common ancestor, and they can develop a family tree.

Australopithecus

Scientific evidence indicates that the first *hominids* (humanlike creatures) belonged to a group called *Australopithecus.* Members of the genus *Australopithecus* are believed to have displayed a critical step in human evolution: the ability to walk upright on two feet.

In 1924, the complete skull of a young hominid was found in a limestone quarry in South Africa. The fossil was unlike any ever seen before. The fossil was named *Australopithecus africanus.* Since then, several other *A. africanus* fossils have been found, showing heavy jaws, low foreheads, and small brain capacities.

In 1974, a team led by Donald Johanson and Tim White found a nearly complete skeleton of a female in the Afar region of Ethiopia. The fossil, which is earlier than *A. africanus,* is widely known as "Lucy," but it is classified as *Australopithecus afarensis.*

In 1977, Mary Leakey and her group made another important discovery: a set of fossil hominid footprints. The footprints, dated at some 4 million years of age, show that the two *Australopithecenes* that made the footprints walked erect.

Members of the *Australopithecus* genus are considered the first hominids but not the first humans. Their brains were small in comparison with human brains, and they had long, monkeylike arms. Other members of the *Australopithecus* genus have been classified as *A. robustus* and *A. boisei.* These different species of *Australopithecus* lived in Africa and are believed to have been primarily plant eaters. Members of the *Australopithecus* group eventually died out about 1 million years ago.

Homo Habilis

Fossils dating back about 2 million years have been found with brain capacities much larger than any *Australopithecus* fossil. On the basis of brain size, these fossils are named *Homo habilis. Homo habilis* is regarded as the first human and the first species of the genus *Homo. Homo habilis* means "handy human." Members of this species were apparently able to use tools, build shelters, and fashion protective clothing. Members of the species appear to have spent much time in trees as well as on the ground, and to have walked erect on two legs. *Homo habilis* is thought to have been the predominant species during the Stone Age, a time during which humans fashioned stone tools. *Homo habilis* eventually became extinct, presumably about 35,000 years ago.

Homo Erectus

The first hominid to leave Africa for Europe and Asia was *Homo erectus.* Evidence suggests that *Homo erectus* replaced *Homo habilis. Homo erectus* was about the size of modern humans and was fully adapted for upright walking. Its brain was much larger than its ancestor's brain, but it had features that separate it from modern humans. The tools of *Homo erectus* were more sophisticated than those of *Homo habilis. Homo erectus* was perhaps the first hunter-gatherer.

Homo erectus was skillful at hunting and butchering animals and is presumed to be the first fire user. Archaeologists believe *Homo erectus* established early cultures and had methods for communicating information to the next generation. The concept of language is believed to have existed in *Homo erectus.*

Homo Sapiens

The earliest fossils of *Homo sapiens* date to about 200,000 years ago. *Homo sapiens* means "intelligent human," and modern humans are classified in this species. *Homo sapiens* is believed to have evolved from *Homo erectus.* The evolution is thought to have taken place in Africa. The earliest fossils of *Homo sapiens* show a gradual change over the last 200,000 years into varieties of *Homo sapiens,* but not new species.

The Neanderthals lived from about 125,000 to about 35,000 years ago in Europe. Neanderthals were short, stocky, and powerfully built. They had a large skull and heavy face with prominent brow ridges, a sloping forehead, and a heavy jaw but a small chin. Their large heads contained larger brains than those of today's humans. Neanderthals made tools of various types and lived in caves and huts.

While Neanderthals were still in existence, the oldest fully modern *Homo sapiens* emerged. This modern variety is called Cro-Magnon, named for a cave in southwestern France where the first fossils were found. Cro-Magnon was somewhat similar to Neanderthals, but they had smaller heads and less prominent faces. Similar to modern humans, Cro-Magnon hunted, displayed culture, and showed the gradual development that led to today's societies. About 10,000 years ago, the first evidence of cities and social structure existed, and by about 5,000 years ago the first great civilizations began to flourish. Today's humans are the descendants of this variation of *Homo sapiens.*

Chapter Check-Out

Questions

1. The first humanlike creatures able to walk upright are called ______.
   a. *Australopithecus*
   b. *Homo erectus*
   c. *Homo habilis*
   d. *Afarensis*

2. ______ is considered to be the first human.
   a. *Homo sapiens*
   b. *Homo habilis*
   c. *Homo erectus*
   d. *Homo afarensis*
3. The first hunter-gatherer is which of the following?
   a. *Homo erectus*
   b. *Homo habilis*
   c. *Homo sapiens*
   d. Neanderthals
4. The concept of language first existed in which of the following populations?
   a. *Homo habilis*
   b. *Homo erectus*
   c. *Homo sapiens*
   d. Cro-Magnon

Answers

1. a
2. b
3. a
4. b

Chapter 15

THE UNITY AND DIVERSITY OF LIFE

Chapter Check-In

❑ Outlining the biological classification system

❑ Understanding the three-domain classification scheme

Earth today is home to more than 8 million different species. This number is constantly changing, however, as new species are discovered at an outstanding rate. Biologists called *taxonomists* have devised a carefully developed scheme to organize these myriad species. In the mid-1700s, Carolus Linnaeus, a Swedish physician and botanist, published several books in which he described thousands of plant and animal species. Linnaeus grouped the species according to their reproductive parts and developed the two-part binomial taxonomy system of categorizing organisms according to genus and species. Linnaeus's work remains valid. It has been combined with the work of Charles Darwin in the field of evolution to form the foundation of modern taxonomy. Darwin's theory of evolution states that all modern species are derived from earlier species and that all organisms, past and present, share a common ancestry. Darwin's theory of evolution, which has become a unifying theme in biology, is the organizing principle of modern taxonomy.

Taxonomists classify organisms in a way that reflects their biological ancestry. Because the ancestral relationships are complex, the taxonomic schemes are also complex and often the subject of revision. Despite their complexity, the taxonomic schemes provide considerable insight into the unity and diversity of life. The term "classification" is synonymous with the word "taxonomy."

Basics of Classification (Taxonomy)

All organisms in the living world are classified and named according to an international system of criteria that dates to the early part of the twentieth century. The rules of classification establish a procedure to be followed when a new species is identified and named. (The rules of classification apply only to formal scientific names, not to common names.)

The scientific name of any organism, called the **binomial name,** has two elements. For example, humans have the binomial name *Homo sapiens.* The name of any species is two words: the name of the genus, followed by the species modifier. For humans, *Homo* is the genus and *sapiens* is the species modifier. The genus name is generally a noun, while the species modifier is an adjective. Thus, *Homo sapiens* literally translates as "human knowing" (or, more simply, "intelligent human," as stated in Chapter 14).

The generally accepted criterion for defining a species is that organisms of the same species interbreed under natural conditions to yield fertile offspring. Individuals of different species normally do not mate. If they are forced to mate, either the mating is unsuccessful or the offspring are sterile. For example, a horse (*Equus caballus*) can be mated to a donkey (*Equus assinus*), and the result will be a mule. However, mules are sterile and cannot reproduce. Thus, the horse and donkey are classified as different species. A quarter horse and a thoroughbred can mate and produce a fertile offspring. Therefore, both are classified as the same species: *Equus caballus.*

For humans, there is only one living species: *Homo sapiens.* However, in past ages, other species, such as *Homo erectus,* may have coexisted with *Homo sapiens. Homo erectus* (see Chapter 14) is considered a separate species because presumably it could not mate with *Homo sapiens.*

The classification scheme provides a mechanism for bringing together various species into progressively larger groups. Taxonomists classify two species together in the same **genus** (the plural is *genera*). For example, the horse *Equus caballus* and the donkey *Equus assinus* are both placed in the genus *Equus.* Similar genera are brought together to form a **family.** Similar families are classified within an **order.** Orders with similar characteristics are grouped in a **class.** Related classes are grouped together as **divisions** or **phyla** (the singular is *phylum*). Divisions are used for plants and fungi, while phyla are used for animals and animal-like organisms. The largest and broadest category used to be the **kingdom,** but this has been usurped by the taxonomic category **domain.**

The classification of a human shows how the classification scheme works. Working from the top down, the human is classified first in the domain Eukarya because it is composed of eukaryotic cells. Next is kingdom Animalia because it has the properties of animals. Animals are then divided into at least 38 phyla, one of which is Chordata. Members of this phylum all have backbones at some time in their lives.

Members of the phylum Chordata are then subdivided into various classes. Humans belong to the class Mammalia, together with other mammals (all of which possess mammary glands and nurse their young). The Mammalia are then divided into several orders, one of which is Primata. Humans belong to the order Primata along with other primates, such as gorillas and monkeys. The order Primata is subdivided into several families, one of which is Hominidae, the family that includes humans and humanlike creatures. Within the family of Hominidae is the genus *Homo,* which includes several species. One of these species is *Homo sapiens.*

Domains and Kingdoms of Living Things

In his classification scheme, Linnaeus recognized only two kingdoms of living things: Animalia and Plantae. At the time, microscopic organisms had not been studied in detail. Either they were placed in a separate category called Chaos or, in some cases, they were classified with plants or animals. Then in the 1860s, the German investigator Ernst Haeckel proposed a three-kingdom system of classification. Haeckel's three kingdoms were Animalia, Plantae, and Protista. Members of the kingdom Protista included the protozoa, fungi, bacteria, and other microorganisms. Haeckel's system was not widely accepted, however, and microorganisms continued to be classified as plants (for example, bacteria and fungi) or animals (for example, protozoa).

In 1968, Robert Whittaker devised a system that was widely accepted by biologists for many years. Whittaker's classification scheme recognized five kingdoms: Monera, Protista, Fungi, Plantae, and Animalia. Even though kingdoms Fungi, Plantae, Animalia, and, to a lesser extent, Protista remain intact today, kingdom Monera is no longer considered a valid category after genetic work shed new light on its phylogenic relatedness.

In the 1970s, genetic comparisons using small-subunit ribosomal RNA highlighted a major problem with grouping all prokaryotic cells together under the single kingdom of **Monera.** Some prokaryotic cells once considered bacteria were more closely related to eukaryotes. This required a larger phylogenic category to be created: the domain. Two of the three domains contain prokaryotic cells (domains Bacteria and Archaea), and all eukaryotic organisms reside under domain Eukarya.

Domain Bacteria includes the vast majority of "everyday prokaryotes," the ones we interact with on a daily basis. This includes pathogenic strains that cause diseases and beneficial species that populate our foods and inhabit our bodies. Domain Archaea, on the other hand, includes those prokaryotes that prefer to live in very extreme environments, reminiscent of early Earth conditions (thus their name, the "archaic" prokaryotes). The third domain includes all life composed of eukaryotic cells. This vast array of life includes kingdoms Protista, Fungi, Plantae, and Animalia.

Kingdom **Protista** has itself been under intense revision, though it is still referred to as its own kingdom. The protists include protozoa, algae, and slime molds. The cells of these organisms are eukaryotic. They can be either unicellular or multicellular, and they may be autotrophic or heterotrophic. Eukaryotic organisms have a nucleus and membrane-bound organelles in their cytoplasm, possess multiple chromosomes, have large ribosomes, and reproduce by mitosis.

The kingdom **Fungi** includes the yeasts, molds, mildews, mushrooms, and other similar organisms. The cells of this kingdom are eukaryotic and heterotrophic. Yeasts are unicellular, whereas other species form long chains of cells and are called *filamentous* fungi. A cell wall strengthened by chitin is found in most members. Food is taken in by the absorption of small molecules from the external environment; thus, fungi are deemed absorptive heterotrophs.

The next kingdom is **Plantae.** Classified here are the mosses, ferns, and seed-producing and flowering plants. All plant cells are eukaryotic and autotrophic. The organisms synthesize their own foods by photosynthesis, and their cell walls contain cellulose. All the organisms are multicellular.

The final kingdom, **Animalia,** includes animals. Animals without backbones (invertebrates) and with backbones (vertebrates) are included here. The cells are eukaryotic; the organisms are heterotrophic. All animals are multicellular, and none has cell walls. In the kingdom Animalia, biologists classify such organisms as sponges, hydras, worms, insects, starfish, reptiles, amphibians, birds, and mammals. The feeding form is one in which large molecules from the external environment are consumed and then broken down to usable parts in the animal body.

Chapter Check-Out

Questions

1. The ideas of Darwin and ______ combined to form the foundation of modern taxonomy.
   a. Linnaeus
   b. Leakey
   c. Mendel
2. The order of the classification system is which of the following?
   a. domain, kingdom, phylum, class, order, family, genus, species
   b. domain, kingdom, phylum, order, class, family, genus, species
   c. domain, kingdom, phylum, class, family, order, genus, species
3. The kingdom ______ contains the yeasts and molds, whereas the kingdom ______ includes the algae and protozoa.
   a. Protista; Fungi
   b. Protista; Plantae
   c. Fungi; Protista

Answers

1. a
2. a
3. c

Chapter 16

PROKARYOTES AND VIRUSES

Chapter Check-In

- ❑ Recognizing the two domains of prokaryotes (Bacteria and Archaea)
- ❑ Understanding the importance of prokaryotes to human health
- ❑ Comparing prokaryotes and the viruses

Prokaryotes are microscopic organisms that include the domains Bacteria and Archaea. **Prokaryotes** lack a nucleus, and they have no organelles except ribosomes. The hereditary material exists as a single loop of double-stranded DNA in a nuclear region, or nucleoid. Prokaryotic cells multiply by an asexual process called *binary fission.* No evidence of mitosis is apparent in the reproductive process.

Prokaryotic cells live in virtually all environments on Earth, including soil, water, and air. They have existed for approximately 3.5 billion years, and they have evolved into every conceivable ecological niche on, above, and below Earth's surface.

Prokaryotic cells are often categorized according to their shapes. The spherical bacteria are referred to as **cocci** (the singular is *coccus*); the rod-shaped bacteria are **bacilli** (the singular is *bacillus*); and the spiral bacteria are **spirochetes** if they are rigid or **spirilla** (the singular is *spirillum*) if they are flexible.

Recent research has focused on the microbiome and the important role prokaryotes play in human health and disease. The **microbiome** is the numerous bacterial communities found on and within the human body. Our intestines alone are home to approximately 700 species of prokaryotes, and their numbers outnumber all human cells in the body tenfold.

Domain Bacteria

Characteristics of bacteria

Domain Bacteria includes the prokaryotes people encounter on an everyday basis. Most bacterial species are **heterotrophic;** that is, they acquire their food from organic matter. The largest number of bacteria are *saprobic,* meaning that they feed on dead or decaying organic matter. A few bacterial species are *parasitic;* these bacteria live within host organisms and cause disease.

Certain bacteria are **autotrophic,** meaning they synthesize their own foods. Such bacteria engage in the process of photosynthesis. They use pigments dissolved in their cytoplasm for the photosynthetic reactions (see Chapter 5). Two groups of *photosynthetic* bacteria are the green sulfur bacteria and the purple bacteria. The pigments in these bacteria resemble plant pigments. Some autotrophic bacteria are *chemosynthetic.* These bacteria use chemical reactions as a source of energy, and synthesize their own foods using this energy.

Bacteria may live at a variety of temperatures. Bacteria living at very cold temperatures are *psychrophilic,* those living at human body temperatures are *mesophilic,* and those living at very high temperatures are *thermophilic.* Bacteria that require oxygen for their metabolism are referred to as **aerobic,** while species that thrive in an oxygen-free environment are said to be **anaerobic.** Some bacteria can live with or without oxygen; they are described as *facultative.* Most bacterial species live in a neutral pH environment (pH of 7), but some bacteria can live in acidic environments (such as in yogurt and sour cream), and others can live in alkaline environments. Certain bacteria are known to live at the pH of 2 found in the human stomach.

Activities of bacteria

Bacteria play many beneficial roles in the environment. For example, some species of bacteria live on the roots of pod-bearing plants (*legumes*) and "fix" nitrogen from the air into organic compounds that are then available to plants. The plants use the nitrogen compounds to make amino acids and proteins, providing them to the animals that consume them. Other bacteria are responsible for the decay that occurs in landfills and other debris in the environment. These bacteria recycle the essential elements in the organic matter.

In the food industry, bacteria are used to prepare products such as cheeses, fermented dairy products, sauerkraut, and pickles. In other industries, bacteria are used to produce antibiotics, chemicals, dyes, numerous vitamins and enzymes, and a number of insecticides. Today, they are used in genetic engineering to synthesize certain pharmaceutical products that cannot be produced otherwise (see Chapter 11).

Cyanobacteria are photosynthetic bacteria formerly known as *blue-green algae.* Most are found in the soil and in freshwater and saltwater environments. The majority of species are unicellular, but some may remain linked and form filaments.

Cyanobacteria, which are autotrophic, serve as important fixers of nitrogen in food chains. In addition, cyanobacteria, a key component of the plankton found in the oceans and seas, produce a major share of the oxygen present in the atmosphere, while also serving as food for fish. Some species of cyanobacteria coexist with fungi to form **lichens.**

Cyanobacteria have played an important role in the development of Earth. Scientists believe that they were the first photosynthetic organisms to occur on Earth's surface. Beginning about 2.7 billion years ago, the oxygen produced by cyanobacteria enriched Earth's atmosphere and converted it to its modern form. This conversion made possible all life forms that use oxygen for cellular respiration.

In the human intestine, bacteria play an important role in our digestive health. Some species synthesize several vitamins not widely obtained in food, especially vitamin K. Bacteria also often break down certain foods that otherwise escape digestion in the body. Signals from gut microbes spur capillary development in the small intestine, increasing absorption of nutrients into the bloodstream.

Unfortunately, many bacteria are **pathogenic;** that is, they cause human disease. Such diseases as cholera, tuberculosis, gonorrhea, syphilis, scarlet fever, food poisoning, Lyme disease, plague, tetanus, typhoid fever, and most pneumonias are due to bacteria. In many cases, the bacteria produce powerful toxins that interfere with normal body functions and bring about disease. Botulism (food poisoning) and tetanus toxins are examples. In other cases, bacteria grow aggressively in the tissues (for example, tuberculosis and typhoid fever), destroying them and thereby causing disease.

Domain Archaea

Characteristics of archaea

Inhabitants of domain Archaea are more closely related to eukaryotic cells than they are to bacteria. Whereas both bacteria and archaea lack a nuclear envelope and membrane-bound organelles, archaea and eukaryotes have similarities beyond those seen between bacteria and eukaryotic cells. Bacteria have the macromolecule peptidoglycan in their cell walls; archaea and eukarya lack this polymer. Bacteria have only a single kind of RNA polymerase, whereas archaea and eukaryotes have several kinds. Introns and DNA-associated histone proteins—a defining characteristic of eukaryotic cells—are absent (histones) or rarely seen (introns) in bacteria but are present in some archaea species.

Activities of archaea

Domain Archaea contains some prokaryotic cells that live in conditions that are too extreme for other forms of life. **Thermophiles** thrive at temperatures as hot as 90°C (194°F) and are found in deep-sea volcanic vents and hot springs. These prokaryotes have evolved to withstand temperatures that would otherwise kill cells by denaturing DNA and proteins. The biotechnological method of polymerase chain reaction (PCR) is possible because of the heat-tolerant DNA polymerase isolated from a thermophile. The extreme **halophiles** withstand highly saline conditions such as the Great Salt Lake and the Dead Sea.

Not all archaea are extremophiles, however; many live in more moderate environments. **Methanogens** are found in many diverse niches, including swamps, termite and cow guts, and deep-sea hydrothermal vents. Methanogens are strict anaerobes that cannot tolerate oxygen. They generate energy using CO_2 to oxidize H_2, creating CH_4 (methane) as a waste product. Methanogens play an important role in water remediation in sewage treatment facilities due to their ability to anaerobically decompose organic waste matter.

Viruses

Technically, viruses are not members of any domain of life. They are considered here because, like bacteria, they are microscopic and can cause human diseases. **Viruses** are acellular particles that lack the properties of living things but have the ability to replicate inside living cells. They have no energy metabolism, they do not grow, they produce no waste products, they do not respond to stimuli, and they do not reproduce independently. In the view of biologists, they are not alive.

Viruses consist of a central core of either DNA or RNA surrounded by a coating of protein. The core of the virus that contains the genes is the **genome,** while the protein coating is the *capsid.* Viruses have characteristic shapes. Certain viruses have the shape of an *icosahedron,* a 20-sided figure made up of equilateral triangles. Other viruses have the shape of a *helix,* a coil-like structure. The viruses that cause herpes simplex, infectious mononucleosis, and chickenpox are *icosahedral* (solid with 20 faces). The viruses that cause rabies, measles, and influenza are *helical* (spiral-shaped).

Viruses reproduce only within living cells. They attach to the plasma membrane of the host cell and release their nucleic acid into the cytoplasm of the cell. The capsid may remain outside the cell, or it may be digested by the host cell within the cytoplasm. In the host cytoplasm, the DNA or RNA of the viral genome encodes the proteins that act as enzymes for the synthesis of new viruses. The enzymes use amino acids in the cell for protein synthesis and nucleotides from the host DNA for nucleic acid synthesis. The viruses obtain cellular ATP and use cellular ribosomes for additional viral synthesis. After some minutes or hours, the new viral capsids and genomes combine to form new viruses.

Once formed, the viruses may escape the host cell when the host cell disintegrates. Alternately, the new viruses may force their way through the plasma membrane of the cell and assume a portion of the plasma membrane as a viral envelope. In either process, the cell is often destroyed and hundreds of new viruses are produced.

Viruses can cause a number of human diseases, including measles, mumps, chickenpox, AIDS, influenza, hepatitis, polio, and encephalitis. Protection from these diseases can be rendered by using vaccines composed of weak or inactive viruses. A viral vaccine induces the immune system to produce antibodies, which provide long-term protection against a viral disease.

Chapter Check-Out

Questions

1. Which of the following is not a characteristic of prokaryotes?
   - **a.** single-celled
   - **b.** contain circular DNA
   - **c.** reproduction by binary fission
   - **d.** contain membrane-bound organelles
2. Bacteria that are able to live both with and without oxygen are called ______.
   - **a.** thermophilic
   - **b.** facultative
   - **c.** aerobic
   - **d.** anaerobic
3. True or False: Prokaryotes contain the hereditary material in the nucleus of the cell.
4. Viruses are known to cause all of the following diseases EXCEPT ______.
   - **a.** chlamydia
   - **b.** influenza
   - **c.** rabies
   - **d.** infectious mononucleosis
5. True or False: Viruses are members of the domain Bacteria.

Answers

1. d
2. b
3. False
4. a
5. False

Chapter 17
PROTISTA

Chapter Check-In

- ❑ Understanding the types of protozoa
- ❑ Getting to know the slime molds
- ❑ Simplifying the divisions of algae

Further analysis of kingdom Protista has suggested the need for restructuring phylogenic classification. Genetic and morphological research has led to subdividing the Protista kingdom into numerous separate kingdoms, each with its own lineage of protist. In truth, the classification of the protists remains in flux. This chapter provides descriptions of representative members of the Protista group and their roles in our lives and the environment, using the still-accepted kingdom and phyla nomenclature.

Members of the kingdom Protista are a highly varied group of organisms, all of which are eukaryotic. In addition, protists are unicellular or, in some cases, colonial. Many species are **autotrophs,** creating their own food, while others are **heterotrophs,** feeding on organic matter. Many species are nonmotile, but the majority of protists are able to move by various means. Many protists have contractile vacuoles, which help them to remove excessive amounts of water from their cytoplasm. The kingdom Protista includes the protozoa, slime molds, and algae.

Protozoa

Protozoa can be divided into four phyla based on their locomotion: Mastigophora, Sarcodina, Ciliophora, and Sporozoa.

Mastigophora

Members of the phylum Mastigophora move about by using one or more whiplike flagella. The genus *Euglena* contains flagellated species. Members are freshwater protists with typical eukaryotic properties, including two flagella, reproduction by mitosis, and flexible nutritional requirements. *Euglena* species also possess chlorophyll within chloroplasts. This pigment allows the organisms to synthesize organic compounds in the presence of sunlight. When no sunlight is available, the organism feeds on dead organic matter in the surrounding environment. Thus, the organism is autotrophic and heterotrophic. Some biologists consider *Euglena* to be the basic stock of evolution for both animals and plants.

Certain species of Mastigophora are zooflagellates, while some are phytoflagellates. The *zooflagellates* live within the bodies of animals and are typified by the wood-digesting flagellates in the intestines of termites. Among the pathogenic zooflagellates are those that cause sleeping sickness, trichomoniasis, and giardiasis. The *phytoflagellates* have photosynthetic abilities and are often discussed with algae in textbooks.

Some species of Mastigophora organize themselves into colonies. Members of the genus *Volvox* are typical colonial forms. The cell colonies are not differentiated into tissues or organs, but the colonies show how a preliminary step in evolutionary development might have occurred.

Sarcodina

Members of the phylum Sarcodina are the **amoebas** and their relatives. Amoebas consist of a single cell without a definite shape. They feed on small organisms and particles of organic matter, and they engulf the particles by phagocytosis. Extensions of the cytoplasm called *pseudopodia* (the singular is *pseudopodium*) assist phagocytosis and motion in the organisms.

Amoebas are found in most lakes, ponds, and other bodies of freshwater. They move by a creeping form of locomotion called *amoeboid motion.* One amoeba called *Entamoeba histolytica* causes a type of dysentery in humans.

Two interesting amoebas are the foraminiferans and the radiolarians. Both are marine amoebas that secrete shells. Their shells have been identified as markers for oil deposits because both were present in the ocean communities that became the organic deposits that, under pressure, became oil fields.

Ciliophora

Members of the phylum Ciliophora move by means of cilia. The organisms are all heterotrophic and have specialized organelles in their cytoplasm. For example, they have two nuclei: a large macronucleus and a number of smaller micronuclei. The micronuclei carry the genetic information of the cell.

The ciliate *Paramecium* typifies the phylum Ciliophora. This organism has a slipper-shaped body with a covering called a *pellicle.* Defensive organelles called *trichocysts* are present in the pellicle. The organism reproduces by mitosis and by an elaborate form of sexual behavior called *conjugation,* which occurs when two *Paramecium* join to one another in the oral region and exchange nuclear material. The cilia of *Paramecium* provide a precise form of motion not provided by flagella or pseudopodia. The cilia can propel the *Paramecium* either forward or backward and move it in a spiral manner.

Sporozoa

Members of the phylum Sporozoa are exclusively parasites. They are so named because some members produce sporelike bodies. Often they have an amoeboid body form, but they are not related to the Sarcodina.

Sporozoans are parasitic organisms with complex life cycles involving several stages. One of the best-known members of the group is the *Plasmodium* species, which are the agents of malaria. The organisms spend portions of their life cycle within mosquitoes. After being injected into the human bloodstream by the mosquito, the parasites invade the red blood cells, undergo numerous changes, and emerge from the red blood cells, destroying them. The infected human experiences a malaria attack soon after.

Another important member of the Sporozoa group is *Toxoplasma gondii,* which causes toxoplasmosis, a disease of the white blood cells. Toxoplasmosis is normally not a dangerous disease, but pregnant women can pass it to their fetuses, where it can cause tissue damage. Also, people who are immunocompromised are susceptible to severe cases of toxoplasmosis.

Two other important pathogens of the Sporozoa group are *Cryptosporidium coccidi* and *Pneumocystis carinii.* Both of these organisms cause severe opportunistic disease in immunocompromised patients. The first causes a severe intestinal diarrhea that is accompanied by the loss of substantial volumes of water; the second causes a type of pneumonia.

Oomycetes

Oomycetes are called *water molds,* a reference to the fact that most species are aquatic. They are not, however, members of kingdom Fungi. During sexual reproduction, the members form clusters of egglike bodies at the tips of their hyphae. Nearby hyphae grow toward the bodies and fuse with them. Nuclear fusions lead to the formation of sexual spores called *oospores* that germinate to produce new hyphae.

In the sexual process of reproduction, oomycetes form a unique cell called a *zoospore.* The zoospore has flagella and is able to move like an animal cell. Certain oomycetes cause downy mildew of grapes, white rust of cabbage, and the late blight of potatoes. Aquatic oomycetes infect fish in aquaria and nature.

Slime Molds

Although slime molds have properties that resemble fungi, many scientists classify the organisms with the protists because of their protozoalike qualities. Slime molds may be **plasmodial slime molds,** which consist of a single, flat, very large cell with many nuclei; or they may be **"cellular" slime molds,** amoebalike cells that live independently and unite with other cellular slime molds to form a single, large, flat cell with many nuclei.

Though scientists are uncertain how slime molds evolved, the organisms show a hint of cellular cooperation that is characteristic of more complex, multicellular organisms. Cellular slime molds normally move about like amoebas. The organisms exist as a mass of cytoplasm with **diploid nuclei.** This mass is a *plasmodium.* Cytoplasmic streaming can be seen within the plasmodium. Slime molds are excellent research tools because they are large and easy to cultivate.

As long as a food supply is adequate and other environmental conditions are optimal, a slime mold grows indefinitely in its plasmodial stage. However, if conditions become harsh, the plasmodium can join with other plasmodia to form a large, sluglike mass. This giant plasmodium can transform itself into a spore-bearing structure similar to a fungus. A stalk arises from the plasmodium, and nuclei within the stalk cells divide by meiosis. Knobs called *sporangia* develop at the end of the stalk and fill with haploid spores. The spores are shed and removed by the wind when the stalk dries. Eventually, the spores germinate into flagellated swarm cells. Swarm cells function as gametes and fuse to form a diploid zygote, which divides by mitosis and forms a new plasmodium. The plasmodium represents the new generation of slime mold.

Algae

The term **algae** refers to a large number of photosynthetic organisms that are multi- or unicellular eukaryotes and are not classified with plants. The organisms are plantlike, however, because they contain chloroplasts with chlorophyll. Most algae can be found in the oceans, but freshwater forms are also abundant.

The algae are subdivided into several divisions (rather than phyla, like the protozoa): Rhodophyta, Pyrrophyta, Chrysophyta, Phaeophyta, and Chlorophyta. The divisions are based in part on the types of pigments and colors they have.

Rhodophyta

Rhodophyta is the division of red algae. These organisms are almost exclusively marine types. Most are unicellular, but some multicellular forms grow anchored to rocks below the level of the low tide. Some are large enough to be seaweeds. Red algae carry on photosynthesis using chlorophyll *a.* The red pigments are very similar to those in many species of cyanobacteria. A derivative of red algae called **agar** is commonly used in bacteriological media in the laboratory.

Pyrrophyta

Members of the Pyrrophyta are dinoflagellates. *Dinoflagellates* are unicellular organisms that are usually surrounded by thick plates that give them an armored appearance. Two flagella move the organism. Many dinoflagellates are luminescent. When affected by sudden movements, they give off light. When optimal conditions exist in the oceans, the dinoflagellates reproduce at explosive rates. Their red pigments cause the water to turn the color of blood. This condition is known as *red tide.*

Chrysophyta

Members of the division Chrysophyta are golden algae, most of which are diatoms. *Diatoms* have cell walls or shells composed of two overlapping halves impregnated with silica. In the oceans, the diatoms carry on photosynthesis. They serve as an important source of food in the oceanic food chains. Diatomaceous earth, a light-colored porous rock composed of the shells of diatoms, is made into a commercial product called *diatomite.* Diatomite is used as a filler, as an absorbent, and as a filtering agent.

Phaeophyta

Members of the division Phaeophyta are the brown algae. These organisms, which are multicellular, are found almost exclusively in saltwater, where they are known as rock weeds and kelp. Despite their great size, the tissue organization in these algae is quite simple compared with plants. Often they are used as fertilizers and sources of iodine.

Chlorophyta

Members of the division Chlorophyta are green algae, some of which are flagellated. The common laboratory specimen *Spirogyra* is classified here, as are other flagellated species. Many colonial forms are also classified as Chlorophyta. The colonial forms may represent the first evidence of multicellularity in evolution, and possess features that are intermediate between those of single-celled eukaryotes and those of more complex multicellular organisms. *Ulva,* the common sea lettuce, is classified here.

Chapter Check-Out

Questions

1. The protozoa are classified into all of the following phyla EXCEPT ______.
   a. Ciliophora
   b. Pyrophyta
   c. Sporozoa
2. The amoebas are members of the phylum ______, whereas the *Paramecium* typifies the phylum ______.
   a. Sporozoa; Ciliophora
   b. Mastigophora; Sarcodina
   c. Sarcodina; Ciliophora
3. True or False: Members of the phylum Sporozoa are exclusively parasites.

Answers

1. b
2. c
3. True

Chapter 18
FUNGI

Chapter Check-In

- ❑ Classing up the kingdom Fungi
- ❑ Tracing sources of fungal disease
- ❑ Introducing the lichens

Fungi, together with bacteria, are the major decomposers of organic matter on Earth. Most fungi are *saprobes;* that is, they digest nonliving organic matter, such as wood, fallen leaves, and dead animals. However, some fungi are parasites that attack living things and cause disease. Fungi cause many agricultural diseases, as well as several human diseases.

A unique physical structure and the method by which they obtain nutrients distinguish fungi from the other kingdoms. Fungi secrete enzymes into the environment and break down organic matter, and then absorb the small nutrient particles through their cell membranes. This process is called *extracellular digestion.*

Except for unicellular yeast, fungi are multicellular organisms, and the body of a fungus consists primarily of cells joined in filaments. Each microscopic filament of a fungus is a *hypha* (the plural is *hyphae*). Hyphae may form a huge tangled interwoven network called a *mycelium* (the plural is *mycelia*). A mycelium is a visible structure. Fungi cells are unique because they possess a polysaccharide called *chitin.* In some cases, the cell walls also contain cellulose. Fungi live in environments that are generally acidic, and they prefer carbohydrate-rich foods.

Reproduction in fungi can be sexual or asexual. In some cases, filaments break from the main mycelium and grow into new individuals. Alternately, a fungus may produce spores by an asexual or sexual process. The spores disperse, germinate, divide, and produce genetically identical fungi. The hyphae and spores are usually *haploid* (containing one set of chromosomes), and the haploid phase dominates the life cycle of a fungus. Spores can withstand extreme dryness and cold to produce a new fungus when conditions permit.

Fungi can also reproduce by a sexual process, which results in a short-lived diploid cell (with two sets of chromosomes) that soon produces haploid cells through meiosis. The spores develop into cells that divide by mitosis to form a new hypha, and then a new mycelium.

In the most accepted classification scheme, fungi are placed in four major divisions: Zygomycetes, Ascomycetes, Basidiomycetes, and Deuteromycetes.

Zygomycetes

The Zygomycetes division of fungi includes mostly terrestrial fungi. Because the hyphae have no cross walls between the cells, they are said to be *coenocytic.* Sexual reproduction in these organisms occurs when sexually opposite hyphae fuse and form spores called *zygospores.*

A common member is the bread mold *Rhizopus stolonifer.* This fungus forms a white or gray mycelium on bread. The sporangia containing the asexual spores can be seen extending into the air. One species of *Rhizopus* is used to ferment rice to make sake, and another species is used in the production of cortisone.

Ascomycetes

Members of the division Ascomycetes are diverse. They range from unicellular yeasts to powdery mildews, cottony molds, and large, complex, "cup" fungi. In the latter, the hyphae are tightly packed and form a cup-shaped structure.

During sexual reproduction, the ascomycetes form a sac called an *ascus.* The ascus forms where sexually opposite hyphae have fused. Within the sac, a number of ascospores form, and each ascospore can reproduce the entire fungus.

Within the Ascomycetes division is the yeast *Saccharomyces.* This organism is used in fermentation processes and baking. The producer of penicillin, *Penicillium,* is also in this class, as is *Aspergillus,* a producer of citric acid, soy sauce, and vinegar. The chestnut blight and Dutch elm diseases are caused by ascomycetes.

Basidiomycetes

Members of the division Basidiomycetes are known as *club fungi.* They include the common mushroom, the shelf fungi, puffballs, and other fleshy fungi. Sexual spores called *basidiospores* are formed on clublike structures called *basidia* (the singular is *basidium*).

Perhaps the most familiar member of this division is an edible mushroom. The mycelium forms below ground. After the hyphae have fused, a mushroom cap emerges. Basidia form on the cap's underside along the gills, and basidiospores form on the basidia. Some mushrooms are edible while others are poisonous, but they are similar in their form and shape.

Basidiomycetes also cause agricultural diseases, including rust and smut diseases. These diseases affect corn, blackberries, and grains such as wheat, oats, and rye.

Deuteromycetes

Those fungi that lack a known sexual reproduction cycle are in the division Deuteromycetes. As far as we know, these fungi reproduce only by an asexual process. (When such an organism's sexual stage is discovered, the fungus is usually reclassified.)

Many human pathogens are currently in this division. These organisms reproduce by fragmenting, with hyphae segments commonly blown about in dust by air currents and the wind. A familiar deuteromycete is the athlete's-foot fungus, which can be picked up from fragments on towels and shower-room floors.

Fungal Disease

Several fungi cause diseases in humans, some of them serious. For example, a disease of the lungs and spinal cord is caused by a fungus called *Cryptococcus neoformans.* This disease is serious in AIDS patients, and it is often spread in dust by air currents.

Another human pathogen is *Candida albicans.* This organism causes disease of the oral cavity (thrush), as well as yeast disease of the reproductive tract. Normally the disease is mild, but in people infected with HIV, for example, it can be serious.

Other human fungal diseases include ringworm and athlete's foot. Each is caused by fungi of various genera, and each is characterized by blister-like regions on the skin or in the webs of toes or fingers. Fungal diseases of the lung tissues include histoplasmosis, blastomycosis, and coccidioidomycosis.

Lichens

Lichens are associations between fungi and cyanobacteria (formerly called blue-green algae). The cyanobacteria are the photosynthetic elements in the association, which is an example of **mutualism** (a living arrangement where both partners benefit). The photosynthetic organisms provide nutrients for themselves and the fungus, while the fungus provides protection, water, and minerals for the photosynthetic metabolism.

Lichens can exist in harsh habitats with extreme climatic conditions. They are found on mountaintops, on rock faces in the desert, and on tree bark. Commonly, lichens set the stage for the growth of more complex plants. Lichens can recover from freezing and drying to resume normal metabolic activities. Their growth is so slow that some patches of lichens may be thousands of years old.

Chapter Check-Out

Questions

1. Each of the following terms is a division of fungi EXCEPT ______.
   a. Zygomycetes
   b. Ascomycetes
   c. Deuteromycetes
   d. Lichens
2. The producer of penicillin, *Penicillium,* is a member of which of the following classes of fungi?
   a. Zygomycetes
   b. Ascomycetes
   c. Deuteromycetes
   d. Basidiomycetes

3. Lichens are associations between fungi and cyanobacteria in which both partners benefit. This arrangement is commonly known as ______.

   a. mutualism
   b. mycelium
   c. chitin
   d. basidia

Answers

1. d
2. b
3. a

Chapter 19

PLANTS: DIVERSITY AND REPRODUCTION

Chapter Check-In

- ❑ Distinguishing characteristics of plants
- ❑ Introducing the nonvascular plants
- ❑ Understanding the vascular architecture

Plants are multicellular eukaryotic organisms with the ability to produce their own food by the process of photosynthesis. (They are autotrophs.) Algae have historically been included with the plants, but they are now classified with the protists. The modern definition of plants includes organisms that live primarily on land (and sometimes in water), excluding algae that live primarily in water.

Another distinguishing characteristic of plants is their type of chlorophyll. **Chlorophyll** is used to absorb energy from the sun during the process of photosynthesis. Plants have chlorophyll *a* and chlorophyll *b,* while many species of algae do not have chlorophyll *b.* Many evolutionary biologists believe that the green algae gave rise to the land plants.

Plants occur in two major groups: nonvascular plants and vascular plants. **Nonvascular plants** do not have specialized tissues to transport fluids, while **vascular plants** do have specialized tissues. The bryophytes (the mosses and liverworts) are the only major group of nonvascular plants. There are three large groups of vascular plants: the **seedless vascular plants** (for example, ferns), the **vascular plants with unprotected seeds** (for example, pines), and the **vascular plants with protected seeds** (for example, flowering plants). While animals are classified in phyla, plants are classified in divisions.

The life cycle of plants has both a multicellular haploid and multicellular diploid phase. Because both phases of the life cycle are multicellular, this type of life cycle is an *alternation of generations.* In contrast, animal life cycles have a multicellular diploid phase and a unicellular haploid phase.

The alternating generations of plants are the sporophyte generation and the gametophyte generation. Individuals in the gametophyte generation (often called *gametophytes*) form gametes, or sex cells. Gametes are haploid cells (they contain one set of chromosomes). Haploid gametes fuse in fertilization. This fusion produces fertilized eggs, which are diploid cells (they have two sets of chromosomes). The plants that develop are diploid plants of the sporophyte generation. Individuals in the sporophyte generation (*sporophytes*) undergo meiosis to produce haploid spores.

Plants produce their gametes in specialized structures. In the nonvascular bryophytes and in the vascular plants, the egg cells are formed in structures called *archegonia* (the singular is *archegonium*). Sperm cells are produced in structures called *antheridia* (the singular is *antheridium*). In some specialized plants, these structures are reduced, and the sporophyte generation is dominant over the gametophyte generation in the life cycle.

Nonvascular Plants

Nonvascular plants belong to the division Bryophyta, which includes mosses, liverworts, and hornworts. These plants have no vascular tissue, so the plants cannot retain water or deliver it to other parts of the plant body. The bryophytes do not possess true roots, stems, or leaves, although the plant body is differentiated into leaflike and stemlike parts. In some species, there are rootlike structures called *rhizoids.* With no vascular tissue, the bryophytes cannot retain water for long periods of time. Consequently, water must be absorbed directly from the surrounding air or another nearby source. This explains the presence of mosses in moist areas, such as swamps and bogs, and on the shaded sides of trees.

The life cycle of the moss is typical of the bryophytes. Flask-shaped archegonia, located among the top leaves of the female gametophytes, produce one egg cell each. Antheridia, located similarly on the male gametophyte, produce many sperm cells that swim in drops of rainwater or dew into the neck of the archegonium to fertilize the egg cell.

The zygote that results from the fertilization develops into a young sporophyte within the archegonium. The sporophyte grows out of the archegonium, taking its nourishment from the gametophyte, and differentiates into a slender stalk with a spore capsule near the tip. Haploid spores are produced by meiosis in this capsule, and when the tip of the capsule opens, the spores are freed. The spores settle in the soil and germinate into gametophytes, which represent the next stage in the alternation of generations.

The life cycles of all bryophytes are uniform, and although the gametophyte generation is the most familiar aspect of the life cycle, neither the sporophyte nor the gametophyte generation is dominant.

Vascular Plants

Vascular plants encompass several divisions of plants and are collectively known as tracheophytes. **Tracheophytes** are characterized primarily by the presence of a vascular system composed of two types of specialized tissue: xylem and phloem. **Xylem** conducts water and minerals upward from the roots of a plant, while **phloem** transports sugars and other nutrients from the leaves to the other parts of the plant. Both xylem and phloem are distributed throughout the plant. The vascular tissue also serves as a means for mechanical support in the plant, so some tracheophytes (such as trees) can grow quite tall.

Seedless vascular plants (ferns)

In many plants, seeds are the structures from which the sporophyte generation emerges. Seeds protect the embryonic plant during its early stages and store food. Many plants do not form seeds in their life cycles, but they have flourished nevertheless. Among the seedless vascular plants are the ferns, classified in the division Pteridophyta.

A mature fern produces spores by meiosis. The spores are stored in cases called *sori* (the singular is *sorus*) on the underside of the fern leaf. The leaf is a *frond,* and the sori resemble dots on the underside of the frond. The fern plant is the sporophyte generation, and this generation dominates the life cycle.

After the spores are dispersed, they germinate into small heart-shaped haploid plants if they reach moist ground. Each plant is called a *prothallus* and is a *gametophyte,* meaning it produces gametes for reproduction. The antheridia and archegonia are found on the underside of the prothallus. Sperm produced in the antheridia swim through moisture on the plant to the archegonia and fertilize the egg cells. The resulting zygote grows within the protection of the archegonium and develops into a young sporophyte. The sporophyte eventually grows out of the archegonium into a slender stalk. The stalk develops into a sporophyte, which, in its adult form, is the familiar fern plant.

Other divisions of the seedless vascular plants are the Psilophyta (the whisk ferns), the Lycophyta (the club mosses), and the Sphenophyta (the horsetails).

Vascular plants with unprotected seeds (gymnosperms)

Vascular plants having unprotected, or naked, seeds are known as **gymnosperms.** Their seeds are not enclosed in female tissues and are therefore said to be naked. There are four divisions of gymnosperms: Cycadophyta (Cycads), Ginkgophyta (Ginkgo), Gnetophyta (Gnetae), and Coniferophyta (Conifer).

Coniferophyta is the largest and most familiar division of the gymnosperms. These plants are cone-bearers and are therefore called *conifers.* Their seeds are borne on the surface of the female cone scales. Members of this division include trees such as cedars, firs, spruces, pines, and giant redwoods. The leaves are generally needle-shaped and contain vascular tissue.

The full-grown conifer (for example, a pine tree) is the sporophyte generation of the plant. The sporophyte produces male and female cones on the same tree. These cones produce spores that undergo meiosis and produce the male and female gametophytes. Male gametophytes are the *pollen grains,* each consisting of four cells. The male gametophyte produces sperm cells in the pollen grains. The female gametophyte produces two or three egg cells that develop within protective structures called **ovules.**

In the spring, the male cone releases pollen, which is blown about by the wind. Some pollen gets trapped on the female cone, where it germinates and forms a pollen tube that makes its way into the ovule. A sperm cell then fertilizes the egg. The zygote that is produced develops into an embryo within the ovule. In time, the embryo matures into a seed. Eventually, the seed falls from the cone and germinates, and the germinating embryo becomes a new pine tree.

The function of dispersal in gymnosperms is assumed by the seeds. Growth of the embryo depends on food supplied from the parent sporophyte, where food-rich tissue surrounds the embryo. The gametophyte generation is little more than a reproductive mechanism in the gymnosperms. Both male and female gametophytes are tiny and entirely dependent on the parent sporophyte.

Vascular plants with protected seeds (angiosperms)

Angiosperms are the most developed and most complex vascular plants. They are the *flowering plants,* of which more than 250,000 species have been identified. Almost all vegetables, flowers, fruits, cereals, grains, grasses, and hardwood trees are angiosperms, the dominant life form on Earth today.

Angiosperm means "seed vessel," a reference to the female tissues that enclose the seed. The tissue is **endosperm.** During embryonic development, the endosperm serves as a source of nourishment. In many angiosperms, the endosperm develops into the fruit of the plant. Thus, the protected seed is often found within a fruit. The two most distinguishing features of angiosperms are the flower and the fruit.

The flower of an angiosperm (see Figure 19-1) consists of a ring of modified leaves called **sepals** that enclose and protect the growing flower bud.

Figure 19-1 The flower of a plant and its structural features.

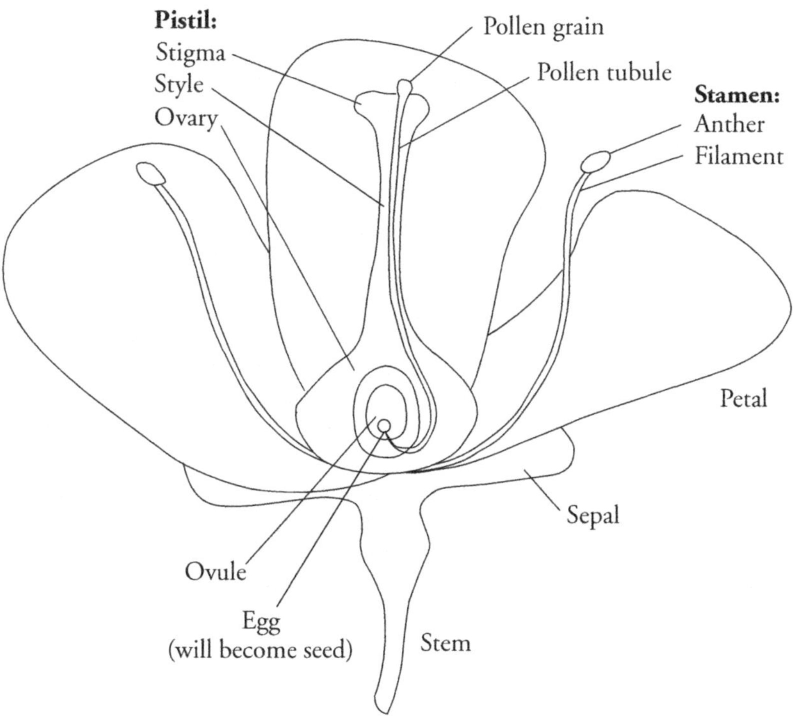

In some species, the sepals are small and green, while in others they become colored and resemble the petals, the next ring. Flower petals are colorful and are useful in attracting pollinating animals, especially insects. Within the petals are the organs of reproduction: the male stamen and, at the center, the female pistil.

The **stamen** of a flower consists of a thin, stemlike filament and an anther, where haploid pollen grains are produced. Each pollen grain is a male gametophyte containing sperm cells. The **pistil** consists of a sticky *stigma;* a *style,* which is a narrow stalk connecting the stigma to the top of the ovary; and the **ovary,** which is where the ovules are enclosed. In each ovule, a single mother cell divides to produce four cells, one of which will develop into a female gametophyte. The female gametophyte is a sac in which there are eight cells, one of which is the egg cell.

In the angiosperm life cycle, a pollen grain lands on the stigma and produces a pollen tube that grows from the pollen grain down through the style, penetrating the ovary to reach the ovule. One of the sperm cells in the pollen tube fertilizes the egg cell in the ovary to produce a diploid **zygote.** The zygote becomes the new plant embryo. Another sperm cell fuses with two other cells in the female gametophyte to produce a triploid (three sets of chromosomes) endosperm nucleus, which develops into the endosperm that will feed the growing embryo. The ovule becomes the seed in which the embryo develops, and the ovary ripens into a fruit.

During plant development, the cells become more specialized. In angiosperms, the embryo remains dormant for a while. When the seed germinates, the embryo grows further. Germination may depend on the availability of oxygen, a suitable temperature, or adequate light.

Germination requires the availability of stored food within the seed. In plants called *monocots* (monocotyledon plants), most of the food is stored in the endosperm in one seed leaf called a *cotyledon* (see Figure 19-2). In *dicots* (dicotyledon plants), most of the food is stored in endosperm in two cotyledons. Among the monocots are grasses, orchids, irises, and lilies. Dicots have two seed leaves (cotyledons) that provide nutrients for the plants as they grow for the seed. Monocots have a single cotyledon. Monocots usually have parallel leaf veins and flower parts occurring in threes. Dicots, in contrast, usually have netlike leaf veins and flower parts in fours or fives. Most of the flowering plants are dicots.

Figure 19-2 Growth of monocotyledon (monocot) and dicotyledon (dicot) plants. Other parts of the young plant are illustrated.

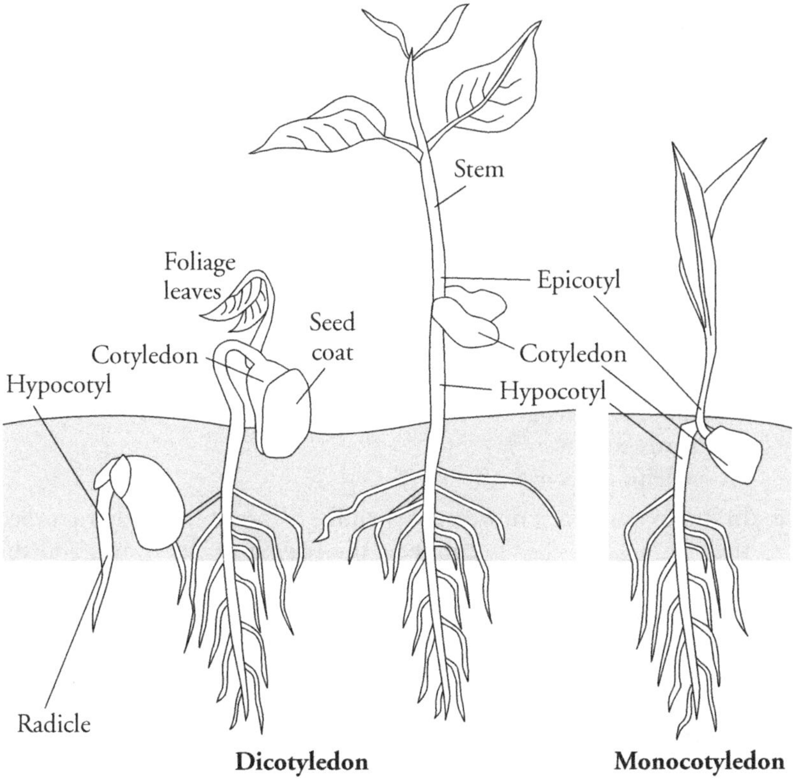

Chapter Check-Out

Questions

1. Each of the following is a group of vascular plants EXCEPT ______.
   a. gymnosperm
   b. angiosperm
   c. bryophytes

2. Conifers (cedars, firs, pines) are members of which of the following groups of vascular plants?
   a. angiosperm
   b. gymnosperm
   c. pteridophyta
3. The pteridophyta produce spores by meiosis, and the spores are stored in cases called ______ on the underside of the leaf, or the ________.
   a. ascus; frond
   b. sori; frond
   c. sorus; xylem
4. The ______ conducts water and transports minerals upward from the roots, whereas the ______ transports sugars and other nutrients from the leaves to the other parts of the plant.
   a. phloem; xylem
   b. sorus; xylem
   c. xylem; phloem
5. In the anatomy of a flower, the stemlike filament and anther is called the ______, a ______ connects the stigma to the ovary, and the ovary is where the ______ are stored.
   a. stamen; style; ovules
   b. stamen; pistil; zygotes
   c. pistil; style; ovules

Answers

1. c
2. b
3. b
4. c
5. a

Chapter 20

VASCULAR PLANTS: STRUCTURE AND FUNCTION

Chapter Check-In

- ❑ Studying the vascular system
- ❑ Differentiating plant tissues
- ❑ Understanding the functions of roots and stems

The plant world is conveniently separated into two major groups: nonvascular plants and vascular plants. The nonvascular plants include the bryophytes, while the vascular plants include the ferns, gymnosperms, and angiosperms (see Chapter 19). The nonvascular plants have no internal transport system. The vascular plants do have such a system, and they are more structurally and functionally complex.

Highly specialized tissues occur in the vascular plants. (A *tissue* is a group of cells working together to carry out a specialized function.) The tissues are organized into specialized organs called roots, stems, and leaves. The internal vascular system, composed of xylem and phloem, is housed within these organs and connects the organs to one another.

Different combinations of tissues make up the organs of a vascular plant. The four types of tissues are as follows: vascular tissue, ground tissue, dermal tissue, and meristematic tissue.

Vascular Tissue

The **vascular tissues** include **xylem,** which conducts water and minerals from the roots upward and throughout the plant, and **phloem,** which transports dissolved nutrients in all directions within the plant.

The main conducting vessels of xylem are the tracheids and the vessels. **Tracheids** are long, thin tubes found in most vascular plants, while **vessels** are large tubes found predominantly in angiosperms. The tracheids and vessels form pipelines that have pores and perforated ends that allow water and minerals to be conducted from one tube to the next and out to the surrounding tissues. Tracheids and vessels also help support the plant body.

The main conducting cells of phloem are sieve cells and sieve tube members. Both cell types have numerous pores through which substances are exchanged with adjacent cells. Sieve tube members occur in angiosperms, while sieve cells are found in other vascular plants. In angiosperms, small cells called *companion cells* assist the sieve tube members in their functions.

Ground Tissue

The **ground tissue** of the vascular plant is responsible for storing the carbohydrates produced by the plant. Ground tissue comprises the majority of a young plant and lies between the vascular and dermal tissues.

The major cells of the ground tissue are *parenchyma cells,* which function in photosynthesis and nutrient storage. They have thin walls, many chloroplasts, and large central vacuoles, and they form the mass of most leaves, stems, and roots.

Another cell of the ground substance is the *sclerenchyma cell.* Sclerenchyma cells are hollow with strong walls, and they help strengthen the ground tissue.

Dermal Tissue

Dermal tissue functions to protect the plant from injury and water loss. Dermal tissue covers the outside of the plant, except in woody shrubs and trees, which have bark. The most common cell type in dermal tissue is the epidermal cell. Generally, a thin, waxy layer called a *cuticle* covers the epidermal cells and protects them. Other cells in the dermal tissue are guard cells that surround the **stomata,** which are openings in the leaves. Gases and water enter and leave the dermal tissue through the stomata.

Meristematic Tissue

Meristematic tissue is growth tissue and the location of most cell division. It is known as *undifferentiated tissue* because cells in the meristematic tissue will eventually become vascular, ground, or dermal tissue.

Plants generally grow where meristematic tissue is present. At the tips of roots and stems, the meristematic tissue is called the *apical meristem.* The primary growth of the plant occurs in the apical meristem.

Lateral growth in a plant is called *secondary growth;* it occurs in *lateral meristem* tissue. Woody trees and shrubs display secondary growth when the plants become enlarged and thickened.

Roots

The primary functions of **roots** in plants are to anchor the plants to the ground and to take in water and minerals from the soil. Substances usually enter the roots by diffusion, but facilitated diffusion may also occur. In addition, roots may be specialized for storage.

Various plants have different types of roots. A *taproot* is a root that grows straight downward and has strong lateral roots growing out of it. Dicots often have taproots. A fibrous root system, consisting of slender, branching roots, occurs in most monocots. Adventitious roots occur on the stems or leaves of some plants, such as corn.

An apical meristem forms the root's tissues. Cells at the tip of the root form the *root cap,* which functions mainly for protection. Behind the root cap is the apical meristem, and behind the apical meristem is the region of elongation, followed by the region of maturation. Many of the epidermal cells of the root develop extensions called *root hairs* that increase the surface area of the root.

Stems

The stems of vascular plants have several functions, including support of the plant, transport of water and minerals by the vascular system, and generation of energy through photosynthetic cells (in some plants). Some stems also function in food and water storage.

Stems arise in the apical meristem. The outer stem layer is the epidermis, the next layer is composed of vascular tissues, the next is the cortex of parenchyma cells, and at the center of the stem is the **pith.**

In herbaceous plants (for example, clover, potatoes, and wheat), the stem is soft and composed primarily of meristematic tissue. In woody plants, in contrast, the stems are hard, with secondary tissues formed after the primary tissues have been laid down. Secondary tissues arise from the *vascular cambium,* a thin layer of dividing cells between the xylem and phloem. The vascular cambium is responsible for the lateral growth in the diameter of the woody stem as the cells lay down secondary xylem toward the inside of the stem and secondary phloem toward the outside of the stem. Xylem becomes the wood of the stem, while phloem, together with a tough tissue called **cork,** becomes the **bark** of the stem. Between the phloem and cork is a thin layer of cells—the *cork cambium*—that produces the cork. Annual rings form from xylem tissue.

Monocots and dicots contrast in the construction of stem tissue. In monocot stems, the vascular bundles are scattered throughout the parenchyma. In dicot stems, the vascular bundles are arranged in a ring around the margin of the stem. Most of the interior dicot stem is occupied by pith. In both monocots and dicots, the xylem provides some structural support for the cell.

In woody stems, the apical meristem is embedded in the tip of the stem within a structure called the *terminal bud.* Along the side of the stem are smaller *lateral buds* from which new branches and twigs emerge. Leaves generally unfold at intervals along the stem as the terminal bud moves upward. The leaves are attached at points on the stem called *nodes.* Spaces between the nodes are called *internodes.* Openings called *lenticels* are found along woody stems. Lenticels function as pores to permit the exchange of gases between the stem tissue and surrounding air.

Leaves

The leaves are the principal organs of photosynthesis in the vascular plants. The cuticle surrounds the epidermis of the leaf to reduce water loss, while gases pass through pores called **stomata.** Beneath the upper epidermis of the leaf is a layer of elongated palisade cells. The *palisade cells* contain numerous chloroplasts, where photosynthesis takes place. Below the palisade cells is the *spongy mesophyll,* an arrangement of loosely packed cells that also contain chloroplasts for photosynthesis. The air spaces around the cells permit efficient gas exchange to take place during photosynthesis.

Bundles of vascular tissues extend through the leaf and form its veins. The vascular tissue supplies water and nutrients to the photosynthetic cells, and the products of photosynthesis are conducted away from the cells through the phloem. Vascular tissue also runs through the *petiole,* the stalk that connects the leaf to the node of the stem. The broad, flat portion of the leaf is the *blade.*

One of the most important activities in the leaf is the opening and closing of the stomata. These pores regulate the rate of gas exchange, which regulates the rate of photosynthesis. The opening and closing of a stoma is regulated by osmotic pressure within a pair of guard cells. Guard cells are thicker on their inner sides than on the outside, so when the cells are swollen with water, they bow outward, opening the stoma. The pressure exerted on the guard cells to open is called **turgor pressure.** Scientists believe that a low concentration of carbon dioxide and an accumulation of potassium ions in the guard cells, along with environmental conditions such as temperature, instigate their opening. ATP provides the energy for opening and closing the guard cells.

When the stomata are open, the carbon dioxide needed for photosynthesis enters the leaf, while the oxygen gas produced in photosynthesis leaves the leaf. The water produced during photosynthesis also leaves through the stomata. This water loss is called *transpiration.*

Water Movement

The movement of water from the roots to the leaves is a critical function in a plant's life. The flow of water depends upon air pressure, humidity, adhesion, and cohesion. At sea level, air pressure can force water up the columns of xylem from the roots to a height of many feet. These columns of water continue to flow upward because water molecules stick to the walls of xylem by *adhesion* and stick to one another by *cohesion.* Water initially moves into the root hair cells by osmosis, because the mineral content of the cells is higher than that of the surrounding environment. Thus, a root pressure is established and extends into the microscopic tubes of the xylem.

At the other end of the plant, water either moves out of the plant through the leaves' stomata or diffuses into the photosynthetic cells and the spaces around them. This loss of water from the xylem at the far end allows water to flow up the xylem tubes, creating continuous water movement.

In the phloem cells, fluids move in a somewhat similar fashion. In many cases, there is a flow of carbohydrates from the source of production to a depository known as the *sink.* For example, sucrose is produced in a photosynthesizing leaf and is transported into phloem tubes. It exits the phloem far away in the root. This is the *source-sink* phenomenon. A constant flow of water is established in the phloem because water follows the carbohydrates by osmosis.

Plant Hormones

The growth and development of many plants are regulated by the activity of **plant hormones. Hormones** are biochemical substances produced in one part of a plant and transported to a different part where they exert a particular effect.

An example of a plant hormone is a series of substances called *auxins.* Auxins increase the length of most plant cells and thereby contribute to the growth and elongation of the plant.

Another plant hormone is *abscisic acid,* which is produced in mature leaves and inhibits growth in developing leaves and germinating seeds. The inhibition occurs during the winter, contributing to the plant's dormancy. Another hormone, *ethylene,* encourages ripening and the dropping of leaves and fruits from the trees. Slight pressure permits the fruit to break loose from the stem.

Two important growth-regulating hormones are the gibberellins and the cytokinins. *Gibberellins* affect plants by stimulating their growth via rapid stem elongation. *Cytokinins* induce plant cells to undergo mitosis; therefore, they encourage increased growth in the roots and stems of plants. They also enhance flowering and stimulate some types of seeds to germinate.

The interactions of hormones and stimuli in the environment often result in a bending or turning response in the plant called a **tropism.** When the plant turns toward a stimulus, the tropism is said to be positive. If the plant turns away from a stimulus, the tropism is negative.

One of the most familiar plant responses is the bending of the stem toward a light source. Light is the stimulus, and the response of the plant is called a **phototropism.** A **geotropism** is a turning of the plant away from or toward the earth. A negative geotropism is a turning away from the earth, such as by a plant stem that grows upward. A positive geotropism is a turning toward the earth, such as in a root that grows downward.

Chapter Check-Out

Questions

1. The main conducting systems of the xylem are the ______ and the ______.
   - a. tracheids; sieve cells
   - b. vessels; sieve cells
   - c. tracheids; vessels
2. The two major cells of the ground tissue are ______ and ______, which function in photosynthesis and nutrient storage.
   - a. parenchyma; sclerenchyma
   - b. parenchyma; epidermal
   - c. epidermal; guard
3. Which of the following is considered to be the principal organ of photosynthesis in the vascular plants?
   - a. stems
   - b. meristem
   - c. leaves
4. The plant hormone ______ inhibits the growth of developing leaves and germinating seeds.
   - a. cytokines
   - b. abscisic acid
   - c. auxin

Answers

1. c
2. a
3. c
4. b

Chapter 21
ANIMALS: INVERTEBRATES

Chapter Check-In

- ❑ Recognizing the sponges
- ❑ Living in marine environments: jellyfish, mollusks, and sea urchins
- ❑ Introducing the flatworms, roundworms, and earthworms
- ❑ Crawling around with the spiders, centipedes, and insects

Animals are multicellular, eukaryotic organisms. They are different from plants in that they take in food and digest it into smaller components. They are heterotrophic rather than autotrophic. In addition to these characteristics, animals are able to move during some point of their life cycle. Also, the primary mode of reproduction in animals is sexual. An animal grows and changes from a single fertilized egg cell into a multicellular organism, passing through various stages of development during its life cycle.

Two major groups of animals exist in the world: invertebrates and vertebrates. **Invertebrates** are animals that have no backbones, while **vertebrates** are animals with backbones (see Chapter 22). Among the invertebrates are numerous phyla of animals comprising approximately 95 percent of all the animal species.

Porifera

The phylum Porifera includes a number of simple animals commonly referred to as *sponges.* Sponges filter and consume fine food particles through their pores. Most sponges live attached to rocks, plants, or other animals in marine environments.

The simplest sponges resemble vases or clusters of tubes with irregular shapes. Each sponge has a large central opening and hundreds of pores in the body wall. Cells lining the cavity remove and digest suspended food particles and water. Waste exits through the central opening.

Sponges are generally without symmetry, and they have a relatively simple cellular organization. The body wall contains a protective layer of flat cells on the outside, an inner layer of flagellated cells, and a gelatinous filling between the two layers. This gelatinous layer contains a group of wandering cells called *amoebocytes.* A sponge is supported by protein that may contain spikes of silicon or calcium compounds. This protein "scaffolding" is typical of the bath sponge.

The cells in the sponge act independently, and there is no evidence of tissue organization. Reproduction occurs by sexual or asexual means. It is possible that sponges evolved from protozoa living in colonies.

Cnidaria

Members of the phylum Cnidaria include hydras, jellyfish, sea corals, and sea anemones. Cnidarians live primarily in marine environments. They have tissue organization and a body plan displaying *radial symmetry;* that is, the organisms are circular with structures that radiate outward. The ends of the structures have tentacles with stinging devices called *cnidocyte* that help in defense and in trapping food. Digestion of food occurs within the central body cavity called the *gastrovascular cavity.* Cells lining this cavity produce enzymes to break down the food, and the products of digestion are taken into the cells. In addition, the animals have a loose network of nerve cells within their tissue. These nerve cells coordinate the animal's activities.

There are two basic body plans in the cnidarians: a hollow, vaselike body called the *polyp,* and an inverted umbrella shape called the *medusa* (the plural is *medusae*). Hydras occur as polyps, while jellyfish appear as medusae. A three-layered body wall makes up the outer surface of the polyp or medusa.

Platyhelminthes

Members of the phylum Platyhelminthes are flatworms, such as the planarian. Grubs and tapeworms are other examples of flatworms. Flatworms display *bilateral symmetry;* that is, the left and right halves of the body are mirror images of one another. Another characteristic of the platyhelminthes is *cephalization,* which means that one end of the animal functions as a head. The "head" contains a mass of nerve cells that acts as a brain and specialized regions for sensing light, chemicals, and pressure.

Flatworms have three distinct layers of tissue, all composed of living cells. They have true organs and organ systems for digestion, movement, excretion, and reproduction. The digestive system consists of a muscular tube with one opening at the mouth. The *excretory system* consists of a network of water-collecting tubules that empty their contents into sacs leading to the exterior.

Movement occurs by the contraction of muscle cells that lie below the epidermis. The contractions are coordinated by signals from a nervous system. Most platyhelminthes have both testes and ovaries, and the organisms pair up to exchange sperm and eggs in fertilization.

Nematoda

Members of the phylum Nematoda are roundworms, and many are microscopic.

Roundworms have many of the same characteristics as flatworms; for example, bilateral symmetry and cephalization. They also have a digestive tract open at both the mouth and the anus. This tract is suspended in a body cavity that is said to be false, so it is called a *pseudocoel.*

Many species of roundworms are free-living organisms that help consume dead plant and animal matter. Several parasitic roundworms cause human diseases, including trichinosis, hookworm disease, and elephantiasis.

Annelida

Members of the phylum Annelida include several classes of segmented worms, among them the worms found in mud and sand, the familiar earthworms, and the leeches. Segmented worms all display bilateral symmetry, cephalization, an open digestive system, segmentation, and a body cavity.

The body cavity is a true body cavity called a *coelom.* This is a fluid-filled space between the innermost cell layer and the outer two cell layers. In this space, the reproductive and digestive organs have evolved into complex structures with complex functions. The digestive system is cushioned in the coelom, so the activities of the digestive system take place without interacting with the inner or outer body walls.

Earthworms and other annelids have numerous segments, each separated from the others by internal partitions. Funnel-shaped excretory units called *nephridia* are located in most of the segments; they remove water and waste. Needed water is reabsorbed, and waste material passes out of the body through pores in the skin. Each segment has longitudinal and circular muscles that contract, compressing fluid to form a water-based skeleton called a **hydrostatic skeleton.**

The annelids have a digestive and circulatory system running their entire length. The circulatory system is closed, and blood is circulated by the contraction of several muscular vessels called *hearts.*

All annelids reproduce sexually.

Mollusca

Members of the phylum Mollusca are soft-bodied animals, such as the snail, clam, squid, oyster, and octopus. Generally, these animals live in water. Some secrete a hard shell. Each mollusk has a muscular organ called a *foot* that is used for gripping or creeping over surfaces. The animal has a head with a mouth; a brain or a sense organ; and groups of internal organs for circulation, excretion, respiration, and reproduction. All mollusks have a *mantle,* a thick fold of tissue that covers the internal organs. The space between the mantle and the body organs is called the *mantle cavity.*

Unlike the previous invertebrates presented in this chapter, mollusks have a well-developed circulatory system with a chambered heart for pumping blood. The blood passes through gills, where oxygen is obtained from water and carbon dioxide is given off. The circulatory system is said to be "open" because blood passes into open spaces at one point during its circulation.

Most mollusks have well-developed nervous systems with large brains and sense organs. Most of these animals have a closed digestive system with only one opening, and a true coelom. All animals of the phylum Mollusca have bilateral symmetry.

Three major classes make up the phylum Mollusca. The first class includes the *gastropods,* a group of land snails and slugs. The second class encompasses the *bivalves,* a group of oysters, clams, and other organisms with two shells. The third class is the *cephalopods,* a group of animals with numerous arms and suckers that extend from the "head foot"; octopuses and squids are in this group.

Arthropoda

The largest number of species in the animal kingdom belongs to the phylum Arthropoda. Members of this phylum, called arthropods, include such animals as spiders, ticks, centipedes, lobsters, and insects. All animals in the phylum have an external skeleton, a segmented body, and jointed appendages.

The external body skeleton of arthropods is called the **exoskeleton.** It surrounds the animal and provides support. The bodies of arthropods are often divided into distinct regions called the head, thorax, and abdomen. Mouthparts exist in the head region. Antennae, the sense organs, are located on the head also, if they are present. Respiration in the arthropods occurs through microscopic holes in the exoskeleton and body wall. Branching networks called tracheae extend from these holes to all parts of an arthropod's body.

Five major classes divide the phylum Arthropoda. The first two classes include the centipedes and millipedes. These animals have multiple appendages and are able to move efficiently on land. Another class of arthropods is the crustaceans, including crabs, shrimp, lobster, and crayfish. In these animals, the exoskeleton, hardened with calcium salts, serves as a protective shell. The arachnids comprise the fourth class of arthropods, including spiders, ticks, mites, and scorpions. Arachnids have a fused head and thorax and have four pairs of legs.

The largest group of arthropods is the insects, including grasshoppers, butterflies, beetles, and cockroaches. Insects live in every conceivable environment on Earth and are among the most highly adapted of all animal species, with well-developed organs for various senses, including smell, touch, taste, and hearing. Insects have three pairs of jointed legs, and many have one or two pairs of wings. The body is divided into three parts: head, thorax, and abdomen. In many cases, the embryo develops into an immature form called the *larva.* The larva transforms into a *pupa* and then becomes an adult. This change is called *metamorphosis.*

Echinodermata

The phylum Echinodermata includes animals that have an internal skeleton and a water-based pressure system for locomotion. Members of this phylum include sea stars, brittle stars, sea urchins, and sea cucumbers. The embryo of echinoderms is different from that of all other invertebrates, and its structure suggests an ancestry to the phylum Chordata.

All echinoderms have spiny skin that helps protect them from predators. All are marine creatures, and all are radially symmetrical as adults. A ring of nerve tissue surrounds the mouth, and nerves branch off from this tissue to extend throughout the body.

Echinoderms have the ability to reproduce by regeneration, a process in which a relatively small piece of the animal can regenerate an entire body. All echinoderms have an internal support system called an **endoskeleton** and a large body cavity containing a set of canals called a *water vascular system.*

Chordata

The final phylum of animals is Chordata. All its members, called **chordates,** have bilateral symmetry, as well as a head, a body cavity, a digestive system, and body segmentation.

In addition, chordates have several unique structures. One such structure is the **notochord.** This is a stiff-yet-flexible rod of tissue extending the length of the animal that provides internal support. A second structure is a hollow **nerve cord** (also called a **spinal cord**) that extends the length of the animal just above the notochord. A third characteristic is the presence of *gill slits:* paired openings from the back of the organism's mouth to the outside.

Not all these characteristics exist in the adult form of the chordates; some exist only in the embryonic form. However, all these characteristics exist at some time in the chordate's life cycle. Most biologists agree that chordates evolved from the echinoderms. The special features of the chordates, such as the notochord, spinal cord, and gill slits, are adaptations to the environment.

The chordates are divided into two major groups: the chordates without backbones (the invertebrates) and the chordates with backbones (the vertebrates). The most primitive chordates are invertebrates. They include the tunicates and lancelets, both of which lack a backbone but have all the chordate traits. Both of these groups somewhat resemble tadpoles and are found clinging to rocks in marine environments. The vertebrate chordates are discussed in Chapter 22.

Chapter Check-Out

Questions

1. The members of the phylum Mollusca contain a muscular organ called a ______ and a thick fold of tissue that covers the internal organs called a ______.

   a. mantle; foot
   b. foot; coelom
   c. foot; mantle
   d. coelom; mantle

2. The external skeleton of the members of the phylum Arthropoda is called the ______, whereas the sense organs are called the ______.
   a. exoskeleton; tracheae
   b. exoskeleton; antennae
   c. segment; antennae
   d. hydrostatic skeleton; coelom
3. All members of which of the following phylum have bilateral symmetry, as well as a head, a body cavity, a digestive system, and body segmentation?
   a. Echinodermata
   b. Arthropoda
   c. Porifera
   d. Chordata
4. Sponges make up the phylum ______, whereas jellyfish and sea anemones are members of the phylum ______.
   a. Cnidaria; Porifera
   b. Porifera; Cnidaria
   c. Porifera; Platyhelminthes
   d. Nematoda; Porifera
5. All echinoderms contain an internal support system called a(n) ______ and a large body cavity with canals called a(n) ______.
   a. endoskeleton; coelom
   b. endoskeleton; water vascular system
   c. coelom; endoskeleton
   d. endoskeleton; pseudocoel
6. Roundworms are members of the phylum ______, whereas the segmented worms are members of the phylum ______.
   a. Annelida; Nematoda
   b. Nematoda; Annelida
   c. Perifora; Nematoda
   d. Nematoda; Perifora

Answers

1. c
2. b
3. d
4. b
5. b
6. b

Chapter 22

ANIMALS: VERTEBRATES

Chapter Check-In

- ❑ Characterizing fishes and amphibians
- ❑ Recognizing reptiles and birds
- ❑ Defining groups of mammals

All members of the animal kingdom can be conveniently divided into invertebrates or vertebrates. Invertebrates are animals that lack a backbone, while vertebrates have a backbone. Invertebrates are found in all the animal phyla (see Chapter 21), including the phylum Chordata.

The phylum Chordata includes animals that have a notochord, a hollow nerve cord, and gill slits. Within this phylum, the vertebrates are classified in the subphylum Vertebrata. There are more than 40,000 living species of vertebrates. Vertebrates are the largest animals ever to inhabit Earth; the dinosaurs were vertebrates, as is the blue whale. The vertebrates are divided into several classes, encompassing fishes, amphibians, reptiles, birds, and mammals.

Fishes

Fishes are aquatic animals with a streamlined shape and a functional tail that allow them to move rapidly through water. Fishes exchange gases with their environment through **gills,** although a few species have lungs that supplement the gills for gas exchange.

One class of fishes called *Agnatha* is made up of species that have no jaws. Lampreys and hagfishes are species within this class. These fishes feed by sucking blood and other body fluids from their prey, usually other fishes.

Another class of fishes is *Placodermi*. The placoderms are now extinct and are recognized as the earliest known fishes with jaws. Placoderms were freshwater bottom-dwellers with a partially bony endoskeleton and bony, armored plates. Another class of fishes is *Chondrichthyes*. These fishes have endoskeletons composed entirely of cartilage. This class includes sharks, skates, and rays. Gas exchanges are made at several prominent, vertical gill openings on either side of the throat.

Bony fishes belong to the class *Osteichthyes*. Most of the familiar fishes living today are bony fishes. They live in the oceans (for example, tuna, mackerel, and herring) and in freshwater (for example, striped bass, trout, and goldfish). Bony fishes prospered during the Devonian Period, which is also called the Age of Fishes. Bony fishes have gills, as well as fleshy pectoral and pelvic fins. The ray-finned fishes are the predominant type of bony fishes living today. These fishes have a *swim bladder,* a gas-filled sac near the gut that permits a fish to change its buoyancy.

Amphibians

Amphibians are animals that live both on land and in water. The members of the class *Amphibia* are believed to have evolved from the lobe-finned fishes about 370 million years ago, taking advantage of the higher concentration of oxygen in air than in water. A fossil called *Tiktaalik,* discovered in 2006, was an amazing model of the intermediate form between early tetrapods and their fish ancestors. *Tiktaalik* had qualities of a fish (fins, gills), but unlike a fish, also possessed a full set of ribs, a neck, and shoulders. The bones of the front fin also had the basic pattern (humerus, radius, ulna, wrist bones) of all limbed animals.

Today's amphibians are represented by frogs, toads, and salamanders. Amphibians live on land and breathe air to meet their oxygen demands. Amphibians are also able to exchange gases through their lungs, their skin, and the inner lining of their mouths. Gas exchange is enhanced by an efficient circulatory system.

Amphibians remain in moist environments or water to avoid dehydration. Amphibians also lay their eggs in water because the eggs would quickly dry out on land. Sperm cells are released into the water, where they fertilize the egg mass. The early-stage tadpoles lead an aquatic existence and later emerge onto land as adult amphibians.

Reptiles

Fossils indicate the earliest reptiles lived about 310 million years ago and resembled today's lizards. In ancient times, the predominant reptiles were dinosaurs. The modern survivors of the Age of Reptiles include lizards, snakes, crocodiles, alligators, turtles, and birds (which are included in the reptile clade). Reptiles belong to the class *Reptilia.*

Reptiles display a number of adaptations that support their life on land. They have a dry, scaly skin that retards water loss. The structural makeup of their limbs provides better support and allows them to move more quickly than any amphibian.

Reproduction in reptiles occurs exclusively on land. The male places sperm into the body of the female, and the embryo develops within an egg, which is laid on dry land. Other reptilian characteristics pertain to the respiratory and circulatory systems. The lungs have a greater surface area than that of the amphibians and permit more air to be inhaled. The circulatory system includes a three-chambered heart that separates oxygen-rich and oxygen-poor blood.

Birds

Though they are also classified as reptiles, birds belong to their own class, *Aves.* Birds have many structural adaptations for flight. For example, the body is streamlined to minimize air resistance, and the endoskeleton bones are light and hollow. Many of the bones are also fused to provide compact strength. To enable flight, birds have feathers, which are lightweight adaptations of reptilian scales. Feathers also insulate against the loss of body heat and water.

Birds are **homeothermic,** meaning they are able to maintain a constant body temperature. The rapid pumping of their four-chambered heart and a high blood-flow rate contribute to this characteristic. Insulating feathers also help maintain a constant body temperature.

Mammals

Members of the class *Mammalia* have hair and nourish their young with milk produced by mammary glands. The presence of body hair or fur helps maintain a constant body temperature in homeothermic mammals.

Several types of mammals exist: the monotremes, marsupials, and placentals. **Monotremes** are egg-laying mammals that produce milk. The duck-billed platypus and the spiny anteater are examples. Both are found in Australia and probably developed during the geographic isolation of that continent.

Marsupials are mammals whose embryos develop within the mother's uterus for a short period of time before birth. After birth, the immature babies crawl into the mother's abdominal pouch, where they complete their development. Animals such as the kangaroo, opossum, and koala bear are marsupials.

Placental mammals include many familiar animals, such as rabbits, deer, dogs, cats, bats, whales, monkeys, and humans. These mammals have a **placenta:** a nutritive connection between the embryo and the mother's uterine wall. Embryos are attached to the placenta, and they complete their development within the mother's uterus.

Mammals have spread to virtually all environments on Earth, ranging from the oceans to the deserts. They live underground, on the ground surface, in trees, and in the air. Mammals have a highly developed nervous system, and many have acute senses of smell, hearing, taste, vision, and touch. Mammals rely on memory and learning to guide their activities. They have been able to develop numerous appropriate responses to different environmental situations. They are considered the most successful animals on Earth today. The human body provides many models for how the hierarchical organization of interacting systems can provide specific functions. Specialized cells combine into specific organs, and multiple organs work together to create an organ system with different functions (e.g., respiration and excretion). The following chapters will address these organ systems.

Chapter Check-Out

Questions

1. Fishes exchange gases with the environment through ______.
   a. gills
   b. jaws
   c. lungs
2. Animals that are capable of living both on land and in water are members of the class ______.
   a. *Reptilia*
   b. *Mammalia*
   c. *Amphibia*

3. The circulatory system of ______ includes a three-chambered heart that separates oxygen-rich and oxygen-poor blood.
   a. amphibians
   b. fishes
   c. reptiles
4. Which of the following is NOT a type of mammal?
   a. marsupials
   b. vertebrata
   c. monotremes

Answers

1. a
2. c
3. c
4. b

Chapter 23
NUTRITION AND DIGESTION

Chapter Check-In

- ❑ Understanding nutritional requirements
- ❑ Introducing the human digestive system

Nutrition refers to the activities by which living things obtain raw materials from the environment and transport them into their cells. The cells metabolize these raw materials and synthesize structural components, enzymes, energy-rich compounds, and other biologically important substances. All the elements and compounds taken into a living thing are *nutrients.*

Animals, including humans, are heterotrophic organisms, and their nutrients consist of preformed organic molecules. The organic molecules rarely come in forms that are readily useful, so animals must process the foods into forms that can be absorbed. This processing is called *digestion.*

Nutrition in Animals

The nutritional requirements of most animals are relatively extensive and complex compared with the simple requirements of plants. The nutrients used by animals include carbohydrates, lipids, nucleic acids, proteins, minerals, and vitamins.

- **Carbohydrates** are the basic source of energy for all animals. Animals obtain their carbohydrates from the external environment (compared with plants, which synthesize carbohydrates by photosynthesis). About one-half to two-thirds of the total calories every animal consumes daily are from carbohydrates. **Glucose** is the carbohydrate most often used as an energy source. This monosaccharide is metabolized during cellular respiration (see Chapter 6), and part of the energy is used to synthesize adenosine triphosphate

(ATP). Other useful carbohydrates are maltose, lactose, sucrose, and starch.

- **Lipids** are used to form cellular and organelle membranes, the sheaths surrounding nerve fibers, and certain hormones. One type of lipid, fats, are extremely useful energy sources.
- **Nucleic acids** are used for the construction of deoxyribonucleic acid (DNA), ribonucleic acid (RNA), and ATP. Animals obtain their nucleic acids from plant and animal tissues, especially from cells that contain nuclei. During digestion, the nucleic acids are broken down into nucleotides, which are absorbed into the cells.
- **Proteins** form the framework of the animal body. Proteins are essential components of the cytoplasm, membranes, and organelles. They are also the major components of muscles, ligaments, and tendons, and they are the essential substances of enzymes. Proteins are composed of 20 kinds of amino acids. Although many amino acids can be synthesized, many others must be supplied in the diet. During digestion, proteins are broken down into their constituent amino acids, which are absorbed into the body.
- Among the **minerals** required by animals are phosphorus, sulfur, potassium, magnesium, and zinc. Animals usually obtain these minerals when they consume plants. **Vitamins** are organic compounds essential in trace amounts to the health of animals. Vitamins can be water soluble or fat soluble. Water-soluble vitamins must be consumed frequently, while fat-soluble vitamins are stored in the liver in fat droplets. Among the many essential vitamins are vitamin A for good vision, vitamin B for substances used in cellular respiration (FAD, NAD, and coenzyme A), and vitamin D to assist calcium absorption in the body.

Animals obtain their nutrients through a broad variety of **feeding patterns.** Sponges, for example, feed on small particles of food that enter their pores. Other aquatic organisms, such as sea cucumbers, wave their tentacles about and trap food on their sticky surfaces. Mollusks, such as clams and oysters, feed by filtering materials through a layer of mucus in their gills. Certain arthropods feed exclusively on fluids.

Some animals feed on food masses, and they usually have organs for seizing, chewing, and consuming food. **Herbivores** are animals that eat only plants, while **carnivores** are animals that eat only other animals. **Omnivores,** which consume both plants and animals, are typified by humans.

Human Digestive System

In the human digestive system, large organic masses are broken down into smaller particles that the body can use as fuel. This is a complex process. The breakdown of the nutrients requires the coordination of several enzymes secreted from specialized cells within the mouth, stomach, intestines, and liver. The major organs or structures that coordinate digestion within the human body include the mouth, esophagus, stomach, small and large intestines, and liver.

Mouth

In the human body, the mouth (oral cavity) is a specialized organ for receiving food and breaking up large organic masses. In the mouth, food is changed mechanically by biting and chewing. Humans have four kinds of teeth: *incisors* are chisel-shaped teeth in the front of the mouth for biting; *canines* are pointed teeth for tearing; and *premolars* and *molars* are flattened, ridged teeth for grinding, pounding, and crushing food.

In the mouth, food is moistened by saliva, a sticky fluid that binds food particles together into a soft mass. Three pairs of **salivary glands**—parotid, submaxillary, and sublingual—secrete saliva into the mouth. The saliva contains an enzyme called *amylase,* which digests starch molecules into smaller molecules of the disaccharide maltose.

During chewing, the tongue moves food about and manipulates it into a mass called a *bolus.* The bolus is pushed back into the pharynx (throat) and is forced through the opening to the esophagus.

Esophagus

The **esophagus** is a thick-walled muscular tube located behind the windpipe that extends through the neck and chest to the stomach. The bolus of food moves through the esophagus by **peristalsis:** a rhythmic series of muscular contractions that propels the bolus along. The contractions are assisted by the pull of gravity.

Stomach

The esophagus joins the stomach at a point just below the diaphragm. A valvelike ring of muscle called the *cardiac sphincter* surrounds the opening to the stomach. The sphincter relaxes as the bolus passes through and then quickly closes.

The *stomach* is an expandable pouch located high in the abdominal cavity. Layers of stomach muscle contract and churn the bolus of food with gastric juices to form a soupy liquid called **chyme.**

The stomach stores food and prepares it for further digestion. In addition, the stomach plays a role in protein digestion. Gastric glands called *chief cells* secrete pepsinogen. Pepsinogen is converted to the enzyme pepsin in the presence of hydrochloric acid. Hydrochloric acid is secreted by *parietal cells* in the stomach lining. The pepsin then digests large proteins into smaller proteins called *peptides.* To protect the stomach lining from the acid, a third type of cell secretes mucus that lines the stomach cavity. An overabundance of acid due to mucus failure may lead to an ulcer.

Small intestine

The soupy mixture called chyme spurts from the stomach through a sphincter into the **small intestine.** An adult's small intestine is about 23 feet long and is divided into three sections: the first 10 to 12 inches form the **duodenum;** the next 10 feet form the **jejunum;** and the final 12 feet form the **ileum.** The inner surface of the small intestine contains numerous fingerlike projections called *villi* (the singular is *villus*). Each villus has projections of cells called *microvilli* to increase the surface area.

Most chemical digestion takes place in the duodenum. In this region, enzymes digest nutrients into simpler forms that can be absorbed. Intestinal enzymes are supplemented by enzymes from the **pancreas,** a large, glandular organ near the stomach. In addition, bile enters the small intestine from the gallbladder to assist in fat digestion.

The enzymes functioning in carbohydrate digestion include amylase (for starch), maltase (for maltose), sucrase (for sucrose), and lactase (for lactose). For fats, the principal enzyme is lipase. Before lipase can act, the large globules of fat must be broken into smaller droplets by bile. *Bile* is a mixture of salts, pigments, and cholesterol that is produced by the liver and stored in the gallbladder, a saclike structure underneath the liver.

Protein digestion is accomplished by several enzymes, including two pancreatic enzymes: trypsin and chymotrypsin. Peptides are broken into smaller peptides, and peptidases reduce the enzymes to amino acids. Nucleases digest nucleic acids into nucleotides in the small intestine also.

Most **absorption** in the small intestine occurs in the jejunum. The products of digestion enter cells of the villi, move across the cells, and enter blood vessels called *capillaries.* Diffusion accounts for the movement of many nutrients, but facilitated diffusion is responsible for the movement of glucose and amino acids. The products of fat digestion pass as small droplets of fat into lacteals, which are branches of the lymphatic system.

Absorption is completed in the final part of the small intestine, the *ileum.* Substances that have not been digested or absorbed then pass into the large intestine.

Large intestine

The small intestine joins the large intestine in the lower-right abdomen of the body. The two organs meet at a blind sac called the **cecum** and a small fingerlike organ called the **appendix.** Evolutionary biologists believe the cecum and appendix are vestiges of larger organs that may have been functional in human ancestors.

The large intestine is also known as the *colon.* It is divided into ascending, transverse, and descending portions, each about 1 foot in length. The colon's chief functions are to absorb water and to store, process, and eliminate the residue following digestion and absorption. The intestinal matter remaining after water has been reclaimed is known as *feces.* Feces consist of nondigested food (such as cellulose), billions of mostly harmless bacteria, bile pigments, and other materials. The feces are stored in the rectum and passed out through the anus to complete the digestion process.

Liver

The **liver** has an important function in processing the products of human digestion. For example, cells of the liver remove excess glucose from the bloodstream and convert the glucose to a polymer called **glycogen** for storage.

The liver also functions in amino acid metabolism. In a process called *deamination,* it converts some amino acids to compounds that can be used in energy metabolism. In doing so, the liver removes the amino groups from amino acids and uses the amino groups to produce urea. Urea is removed from the body in the urine (see Chapter 26). **Fats** are processed into two-carbon units that can enter the **Krebs cycle** for energy metabolism. The liver also stores vitamins and minerals, forms many blood proteins, synthesizes cholesterol, and produces bile for fat digestion.

Chapter Check-Out

Questions

1. The basic source of energy for all animals is ______.
   - a. lipids
   - b. carbohydrates
   - c. proteins
   - d. nucleic acids
2. ______ are animals that eat only plants, whereas ______ are animals that eat only other animals.
   - a. Herbivores; omnivores
   - b. Omnivores; herbivores
   - c. Herbivores; carnivores
   - d. Omnivores; carnivores
3. The ______ is a thick-walled muscular tube located behind the windpipe that extends through the neck and chest to the stomach.
   - a. sphincter
   - b. pancreas
   - c. bolus
   - d. esophagus
4. In the stomach, ______ digests large proteins into smaller peptides.
   - a. pepsinogen
   - b. hydrochloric acid
   - c. pepsin
   - d. trypsin
5. Carbohydrates are digested within the ______, whereas the carbohydrate glucose is stored in the ______.
   - a. large intestine; liver
   - b. small intestine; liver
   - c. liver; small intestine
   - d. liver; large intestine

Answers

1. b
2. c
3. d
4. c
5. b

Chapter 24
GAS EXCHANGE

Chapter Check-In

- ❑ Directing gas exchange in animals
- ❑ Respiring through the nose and pharynx
- ❑ Comparing the trachea and lungs

All living things obtain the energy they need by metabolizing energy-rich compounds, such as carbohydrates and fats. In the majority of organisms, this metabolism takes place by respiration, a process that requires oxygen (see Chapter 6). In the process, carbon dioxide gas is produced and must be removed from the body.

In plant cells, carbon dioxide may appear to be a waste product of respiration, too, but because it is used in photosynthesis (see Chapter 5), carbon dioxide may be considered a by-product. Carbon dioxide must be available to plant cells, and oxygen gas must be removed. Gas exchange is thus an essential process in energy metabolism, and gas exchange is an essential prerequisite to life, because where energy is lacking, life cannot continue.

Mechanisms for Gas Exchange

The basic mechanism of gas exchange is diffusion across a moist membrane. **Diffusion** is the movement of molecules from a region of greater concentration to a region of lesser concentration, in the direction following the concentration gradient. In living systems, the molecules move across cell membranes, which are continuously moistened by fluid.

Simple organisms

Single-celled organisms, such as bacteria and protozoa, are in constant contact with their external environment. Gas exchange occurs by diffusion across their membranes. Even in simple multicellular organisms, such as green algae, their cells may be close to the environment, and gas exchange can occur easily.

In larger organisms, adaptations bring the environment closer to the cells. Liverworts, for example, have numerous air chambers in the internal environment. Sponges and hydras have water-filled central cavities, and planaria have branches of their gastrovascular cavity that connect with all parts of the body.

Plants

Although plants are complex organisms, they exchange their gases with the environment in a rather straightforward way. In aquatic plants, water passes among the tissues and provides the medium for gas exchange. In terrestrial plants, air enters the tissues, and the gases diffuse into the moisture bathing the internal cells.

In the leaf of the plant, an abundant supply of carbon dioxide must be present, and oxygen from photosynthesis must be removed. Gases do not pass through the cuticle of the leaf; they pass through pores called **stomata** in the cuticle and epidermis. Stomata are abundant on the lower surface of the leaf, and they normally open during the day when the rate of photosynthesis is highest. Physiological changes in the surrounding guard cells account for the opening and closing of the stomata (see Chapter 20).

Animals

In animals, gas exchange follows the same general pattern as in plants. Oxygen and carbon dioxide move by diffusion across moist membranes. In simple animals, the exchange occurs directly with the environment. But with complex animals, such as mammals, the exchange occurs between the environment and the blood. The blood then carries oxygen to deeply embedded cells and transports carbon dioxide out to where it can be removed from the body.

Earthworms exchange oxygen and carbon dioxide directly through their skin. The oxygen diffuses into tiny blood vessels in the skin surface, where it combines with the red pigmented protein **hemoglobin.** Hemoglobin binds loosely to oxygen and carries it through the animal's bloodstream. Carbon dioxide is transported back to the skin by the hemoglobin.

Terrestrial arthropods have a series of openings called **spiracles** at the body surface. Spiracles open into tiny air tubes called **tracheae,** which expand into fine branches that extend into all parts of the arthropod body.

Fishes use outward extensions of their body surface called gills for gas exchange. **Gills** are flaps of tissue richly supplied with blood vessels. As a fish swims, it draws water into its mouth and across the gills. Oxygen diffuses out of the water into the blood vessels of the gill, while carbon dioxide leaves the blood vessels and enters the water passing by the gills.

Terrestrial vertebrates such as amphibians, reptiles, birds, and mammals have well-developed respiratory systems with lungs. Frogs swallow air into their lungs, where oxygen diffuses into the blood to join with hemoglobin in the red blood cells. Amphibians can also exchange gases through their skin. Reptiles have folded lungs to provide increased surface area for gas exchange. Rib muscles assist lung expansion and protect the lungs from injury.

Birds have large air spaces called *air sacs* in their lungs. When a bird inhales, its rib cage spreads apart and a partial vacuum is created in the lungs. Air rushes into the lungs and then into the air sacs, where most of the gas exchange occurs. This system is birds' adaptation to the rigors of flight and their extensive metabolic demands.

The lungs of mammals are divided into millions of microscopic air sacs called **alveoli** (the singular is *alveolus*). Each alveolus is surrounded by a rich network of blood vessels for transporting gases. In addition, mammals have a dome-shaped diaphragm that separates the thorax from the abdomen, providing a separate chest cavity for breathing and pumping blood. During inhalation, the diaphragm contracts and flattens to create a partial vacuum in the lungs. The lungs fill with air, and gas exchange follows.

Human Respiratory System

The human respiratory system consists of a complex set of organs and tissues that capture oxygen from the environment and transport the oxygen into the lungs. The organs and tissues that comprise the human respiratory system include the nose, pharynx, trachea, and lungs.

Nose

The respiratory system of humans begins with the nose, where air is conditioned by warming and moistening. Bone partitions separate the nasal cavity into chambers, where air swirls about in currents. Hairs and hairlike cilia trap dust particles and purify the air.

Pharynx

The nasal chambers open into a cavity at the rear of the mouth called the **pharynx** (throat). From the pharynx, two tubes called *Eustachian tubes* open to the middle ear to equalize air pressure there. The pharynx also contains tonsils and *adenoids,* which are pockets of lymphatic tissue used to trap and filter microorganisms.

Trachea

After passing through the pharynx, air passes into the windpipe, or **trachea.** The trachea has a framework of smooth muscle with about 16 to 20 open rings of cartilage shaped like a C. These rings give rigidity to the trachea and ensure that it remains open.

The opening to the trachea is a slitlike structure called the **glottis.** A thin flap of tissue called the **epiglottis** folds over the opening during swallowing and prevents food from entering the trachea. At the upper end of the trachea, several folds of cartilage form the **larynx,** or voice box. In the larynx, flaplike pairs of tissues called *vocal cords* vibrate when a person exhales and produce sounds.

At its lower end, the trachea branches into two large **bronchi** (singular, *bronchus*). These tubes also have smooth muscle and cartilage rings. The bronchi branch into smaller **bronchioles,** forming a bronchial "tree." The bronchioles terminate in the alveoli.

Lungs

Human **lungs** are composed of approximately 300 million alveoli. Red blood cells pass through the capillaries in single file, and oxygen from each alveolus enters the red blood cells and binds to the hemoglobin. In addition, carbon dioxide contained in the plasma and red blood cells leaves the capillaries and enters the alveoli when a breath is taken. Most carbon dioxide reaches the alveoli as bicarbonate ions, and about 25 percent of it is bound loosely to hemoglobin.

When a person inhales, the rib muscles and diaphragm contract, thereby increasing the volume of the chest cavity. This increase leads to reduced air pressure in the chest cavity, and air rushes into the alveoli, forcing them to expand and fill. The lungs passively obtain air from the environment by this process. During exhalation, the rib muscles and diaphragm relax, the chest cavity volume diminishes, and the internal air pressure increases. The compressed air forces the alveoli to close, and air flows out.

The nerve activity that controls breathing arises from impulses transported by nerve fibers passing into the chest cavity and terminating at the rib muscles and diaphragm. These impulses are regulated by the amount of carbon dioxide in the blood: A high carbon dioxide concentration leads to an increased number of nerve impulses and a more rapid breathing rate.

Chapter Check-Out

Questions

1. Gases pass through pores called ______ in plants.
   - a. cuticles
   - b. spiracles
   - c. gills
   - d. stomata
2. Fishes exchange gases using ______, whereas the arthropods use air tubes called ______.
   - a. gills; tracheae
   - b. tracheae; lungs
   - c. gills; lungs
   - d. tracheae; alveoli
3. Mammalian lungs are divided into millions of air sacs called ______.
   - a. tracheae
   - b. alveoli
   - c. bronchioles
   - d. diaphragms

4. Several folds of cartilage form the voice box, which is called the ______.
   a. bronchi
   b. alveoli
   c. larynx
   d. pharynx

Answers

1. d
2. a
3. b
4. c

Chapter 25

BLOOD AND CIRCULATION

Chapter Check-In

- ❑ Understanding circulation in plants and animals
- ❑ Introducing the components of circulation
- ❑ Defending against foreign antigens

Single-celled organisms are in constant contact with their environments, obtaining nutrients and oxygen directly across the cell surface. The same holds true for small and simple plants and animals, such as algae, bryophytes, sponges, cnidarians, and flatworms. Larger and more complex plants and animals require methods for transporting materials to and from cells far removed from the external environment. These organisms have evolved transport systems.

Circulatory Systems

Circulatory systems function to transport materials to and from cells throughout an organism. Organisms ranging from plants to animals have different nutritional requirements. Due to these differences, various species have evolved distinct circulatory processes to assist with their specific transport needs.

Plants

Transport systems are found in vascular plants. Vascular networks provide intercellular communication in terrestrial plants. The systems consist of tubelike connective tissues organized into xylem and phloem. Xylem transports water and minerals in the plants, while phloem transports food materials and hormones (see Chapter 20).

Xylem and phloem tissues are grouped in arrangements called **vascular bundles.** In monocot plants, the vascular bundles are scattered throughout the parenchyma tissue in no particular pattern. In dicot plants, the vascular bundles occur in a circle around a central region of pith. In woody dicot plants, new xylem forms on the inside of the cambium each season; the old xylem forms the annual rings of the plant.

Animals

In animals, the transport system is generally called a **circulatory system** because the blood flows through a circuit. Most animals have one or more organs called *hearts* for pumping blood. The channels through which the blood flows are the **arteries** (which lead away from the heart), the **veins** (which lead to the heart), and the **capillaries** (the microscopic blood vessels between arteries and veins).

In animals such as earthworms, the circulatory system consists of blood, channels, and five pulsating vessels that function as hearts, rushing blood through the vessels to all the earthworm's body parts. Gases bind to hemoglobin in the blood.

Terrestrial arthropods have an open circulatory system. A tubelike heart pumps blood into a dorsal blood vessel, which empties into the arthropod's body cavity, or *hemocoel.* Contractions of the body muscles gradually move blood back toward the animal's heart.

All vertebrates have a single, strong muscular heart to pump the blood. In fishes, the blood accumulates in a thin-walled receiving chamber called the **atrium.** The blood then passes through a valve into a pumping chamber, a **ventricle.** The ventricle contracts and forces blood out to the gills, where gas exchange occurs, and from there to the remainder of the body. Veins bring the blood back to the atrium.

Amphibians, such as frogs, have a three-chambered heart. The heart has a right and a left atrium and a single ventricle. In reptiles, a muscular septum exists between the two sides of the ventricle, creating a primitive four-chambered heart. Birds have a more sophisticated four-chambered heart than reptiles. One ventricle pumps blood to the lungs for gas exchange, while the second pumps oxygen-rich blood to the rest of the body tissues. Mammals also have a four-chambered heart.

Human Circulatory System

The human circulatory system functions to transport blood and oxygen from the lungs to the various tissues of the body. The heart pumps the blood throughout the body. The lymphatic system is an extension of the human circulatory system that includes cell-mediated and antibody-mediated immune systems. The components of the human circulatory system include the heart, blood, red and white blood cells, platelets, and the lymphatic system.

Heart

The human heart is about the size of a clenched fist. It contains four chambers: two atria and two ventricles. Oxygen-poor blood enters the **right atrium** through a major vein called the **vena cava.** The blood passes through the **tricuspid valve** into the **right ventricle.** Next, the blood is pumped through the **pulmonary artery** to the lungs for gas exchange. Oxygen-rich blood returns to the **left atrium** via the **pulmonary vein.** The oxygen-rich blood flows through the **bicuspid (mitral) valve** into the **left ventricle,** from which it is pumped through a major artery, the **aorta.** Two valves called **semilunar valves** are found in the pulmonary artery and aorta.

The ventricles contract about 70 times per minute, which represents a person's pulse rate. Blood pressure, in contrast, is the pressure exerted against the walls of the arteries. Blood pressure is measured by noting the height to which a column of mercury can be pushed by the blood pressing against the arterial walls. A normal blood pressure is a height of 120 millimeters of mercury during heart contraction (*systole*) and a height of 80 millimeters of mercury during heart relaxation (*diastole*). Normal blood pressure is usually expressed as "120 over 80."

Coronary arteries supply the heart muscle with blood. The heart is controlled by nerves that originate on the right side in the upper region of the atrium at the sinoatrial node. This node is called the *pacemaker.* It generates nerve impulses that spread to the atrioventricular node, where the impulses are amplified and spread to other regions of the heart by nerves called **Purkinje fibers.**

Blood

Blood is the medium of transport in the body. The fluid portion of the blood, the **plasma,** is a straw-colored liquid composed primarily of water. Important nutrients, hormones, and clotting proteins, as well as waste products, are transported in the plasma. Red blood cells and white blood cells are also suspended in the plasma. Plasma from which the clotting proteins have been removed is called **serum.**

Red blood cells

Red blood cells are also called **erythrocytes.** These are disk-shaped cells produced in the bone marrow. Red blood cells have no nuclei, and their cytoplasm is filled with hemoglobin.

Hemoglobin is a red-pigmented protein that binds loosely to oxygen atoms and carbon dioxide molecules. It is the mechanism of transport of these substances. (Much carbon dioxide is also transported as bicarbonate ions.) Hemoglobin also binds to carbon monoxide. Unfortunately, this binding is irreversible, so it often leads to carbon monoxide poisoning.

A red blood cell circulates for about 120 days and is then destroyed in the spleen, an organ located near the stomach and composed primarily of lymph node tissue. When the red blood cell is destroyed, its iron component is preserved for reuse in the liver. The remainder of the hemoglobin converts to bilirubin. This amber substance is the chief pigment in human bile, which is produced in the liver.

Red blood cells commonly have immune-stimulating polysaccharides called **antigens** on the surface of their cells. Individuals having the A antigen have blood type A (as well as anti-B antibodies); individuals having the B antigen have blood type B (as well as anti-A antibodies); individuals having the A and B antigens have blood type AB (but no anti-A or anti-B antibodies); and individuals having no antigens have blood type O (as well as anti-A and anti-B antibodies).

White blood cells

White blood cells are referred to as **leukocytes.** They are generally larger than red blood cells and have clearly defined nuclei. They are also produced in the bone marrow and have various functions in the body. Certain white blood cells called **lymphocytes** are essential components of the immune system (discussed later in this chapter). Other cells called **neutrophils** and **monocytes** function primarily as **phagocytes;** that is, they attack and engulf invading microorganisms. About 30 percent of the white blood cells are lymphocytes, about 60 percent are neutrophils, and about 8 percent are monocytes. The remaining white blood cells are **eosinophils** and **basophils.** Their functions are uncertain; however, eosinophils are related to parasitic infections and basophils are believed to function in allergic responses.

Platelets

Platelets are small, disk-shaped blood fragments produced in the bone marrow. They lack nuclei and are much smaller than erythrocytes. Also known technically as **thrombocytes,** they serve as the starting material for **blood clotting.** The platelets adhere to damaged blood vessel walls, and thromboplastin is released from the injured tissue. Thromboplastin, in turn, activates other clotting factors in the blood. Along with calcium ions and other factors, thromboplastin converts the blood protein prothrombin into thrombin. Thrombin then catalyzes the conversion of its blood protein fibrinogen into a protein called *fibrin,* which forms a patchwork mesh at the injury site. As blood cells are trapped in the mesh, a blood clot forms.

Lymphatic system

The **lymphatic system** is an extension of the circulatory system consisting of a fluid known as lymph, capillaries called lymphatic vessels, and structures called lymph nodes. **Lymph** is a watery fluid derived from plasma that has seeped out of the blood system capillaries and mingled with the cells. Instead of returning to the heart through the blood veins, this lymph enters a series of one-way **lymphatic vessels** that return the fluid to the circulatory system. Along the way, the ducts pass through hundreds of tiny, capsulelike bodies called **lymph nodes.** Located in the neck, armpits, and groin, the lymph nodes contain cells that filter the lymph and phagocytize foreign particles.

The **spleen** is composed primarily of lymph node tissue. Lying close to the stomach, the spleen is also the site where red blood cells are destroyed. The spleen serves as a reserve blood supply for the body.

Immune system

The immune system protects the body from foreign substances ("antigens") such as viruses and bacteria. The vertebrate immune system is composed of two lines of defense: nonspecific and specific. Nonspecific defenses against foreign invaders include physical barriers such as skin and mucous membranes. There are also phagocytic cells that circulate around the body and engulf any foreign material. The specific line of defense is based on special immune cells in the body called lymphocytes. It is referred to as "specific" because lymphocytes are able to "remember" the exposure to certain antigens, and once the body is exposed to that same antigen a second time, the subsequent **immune response** is faster and more robust.

The lymph nodes are also the primary sites of the white blood cells called **lymphocytes.** The body has two kinds of lymphocytes: **B-lymphocytes** (B cells) and **T-lymphocytes** (T cells). Both of these cells are stimulated by microorganisms or other foreign materials in the blood. These antigens are picked up by phagocytes and presented to the B and T cells in order to trigger the specific immune response. Once a lymphocyte is triggered to combat a particular pathogen, it undergoes a process of "clonal selection." Each particular B and T cell is produced to recognize a slightly different antigen. Once the foreign invader is recognized, the lymphocyte best suited to combat it creates a vast population of clones. Some clones destroy and remove the initial threat, while some remain at guard for a possible second infection (the "memory" of the specific immune system).

Certain antigens, such as viruses, stimulate the T cells. T cells are specialized to protect the body against intracellular invaders by destroying infected host cells. This process is called **cell-mediated immunity.**

B cells are stimulated primarily by toxins and pathogens in the blood and lymph. On stimulation, the B cells convert to large antibody-producing cells called **plasma cells.** The plasma cells synthesize proteins called **antibodies,** which are released into the circulatory system. The antibodies flow to the antigen site and destroy the microorganisms by chemically reacting with them in a highly specific manner. The reaction encourages phagocytosis, neutralizes many microbial toxins, eliminates the ability of microorganisms to move, and causes them to bind together in large masses. This process is called **antibody-mediated immunity.** After the microorganisms have been removed, another form of B cells—memory B cells—remain in the bloodstream and provide long-term protection to the body. Thus, the body becomes immune to specific disease microorganisms.

Chapter Check-Out

Questions

1. Oxygen-poor blood enters the right atrium of the human heart through the ______ and is pumped through the ______ to the lungs.

   a. vena cava; pulmonary vein
   b. pulmonary vein; vena cava
   c. vena cava; pulmonary artery
   d. pulmonary artery; vena cava

2. The fluid portion of the blood is called ______.
    a. serum
    b. plasma
    c. semilunar
    d. basophil
3. True or False: Red blood cells do not contain a nucleus.
4. Which of the following are NOT types of white blood cells?
    a. neutrophils
    b. eosinophils
    c. lymphocytes
    d. thrombocytes
5. An example of a nonspecific immune response would be
    a. T cells attacking an infected body cell
    b. B cells releasing antibodies to fight off an infection
    c. skin providing a physical barrier against infection
    d. memory B cells remembering a past infection

Answers

1. c
2. b
3. True
4. d
5. c

Chapter 26

EXCRETION AND HOMEOSTASIS

Chapter Check-In

- ❑ Maintaining steady-state equilibrium
- ❑ Detailing the human excretory system
- ❑ Controlling kidney function

The excretion process is concerned with the removal of waste products from the animal body. The relatively stable internal environment of the organism, a concept known as **homeostasis,** is dependent on the process of excretion. Homeostasis means "staying the same." It refers to a steady-state equilibrium that exists internally in a healthy organism or cells, despite changes in the external environment. The excretory system plays a major role in homeostasis.

Excretory Systems

Because one-celled organisms are in constant contact with their environment, they do not need excretory organs. However, multicellular organisms need a mechanism to carry waste products from cells to the external environment. Flatworms, such as planaria, have a series of excretory cells called *flame cells.* Flame cells contain cilia that direct water and metabolic wastes to enter the cells and pass into excretory canals. The excretory canals join with other canals to form excretory tubules. Fluid from the excretory tubules leaves the body through pores.

In earthworms, members of the phylum Annelida, the excretory system consists of structural units called *nephridia* (the singular is *nephridium*). Each nephridium contains a ciliated tunnel that leads to a long, coiled tubule, which leads to a bladderlike sac (a primitive bladder). Fluid moves from the internal environment into the funnel. As fluid passes through the tubule, cells in the tubular lining absorb useful compounds such as glucose, amino acids, and salts. The remaining materials constitute metabolic waste, and they are passed into the bladderlike sac. The sac later opens through a pore in the earthworm's skin, from which the waste products are discharged.

Insects have a series of tubules for excretion called *Malpighian tubules.* Fluid enters at the upper end of the tubules and passes down their entire length. The cells in the tubular walls reabsorb precise amounts of water, salts, and other materials to maintain a delicate balance within the insect tissues. The tubules eventually lead to an insect's intestine, where waste products are removed.

Human Excretory System

The human excretory system functions to remove waste from the human body. This system consists of specialized structures and capillary networks that assist in the excretory process. The human excretory system includes the kidneys and their functional unit, the nephron. The excretory activity of the kidneys is modulated by specialized hormones that regulate the amount of absorption within the nephron.

Kidneys

The human kidneys are the major organs of bodily excretion (see Figure 26-1). They are bean-shaped organs located on either side of the backbone at about the level of the stomach and liver. Blood enters the kidneys through **renal arteries** and leaves through **renal veins.** Tubes called **ureters** carry waste products from the kidneys to the **urinary bladder** for storage or for release.

The product of the kidneys is **urine,** a watery solution of waste products, salts, organic compounds, and two important nitrogen compounds: uric acid and urea. **Uric acid** results from nucleic acid decomposition, and **urea** results from amino acid breakdown in the liver. Both of these nitrogen products can be poisonous to the body and must be removed in the urine.

Nephron

The functional and structural unit of the kidney is the **nephron.** The nephron produces urine and is the primary unit of homeostasis in the body. It is essentially a long tubule with a series of associated blood vessels. The upper end of the tubule is an enlarged cuplike structure called the **Bowman's capsule.** Below the Bowman's capsule, the tubule coils to

Figure 26-1 Details of the human excretory system. Position and allied structures of the kidneys (top). A cross section of the kidney showing the two major portions (left). Details of the nephron, the functional unit of the kidney (right).

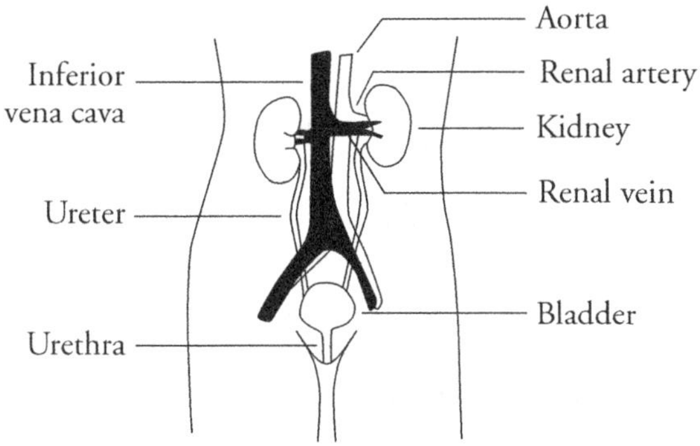

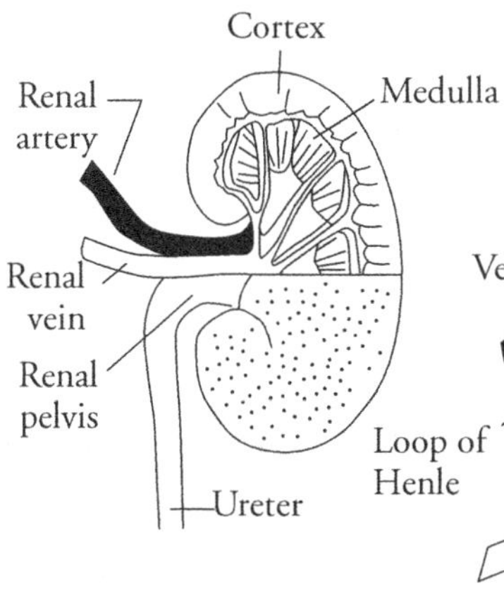

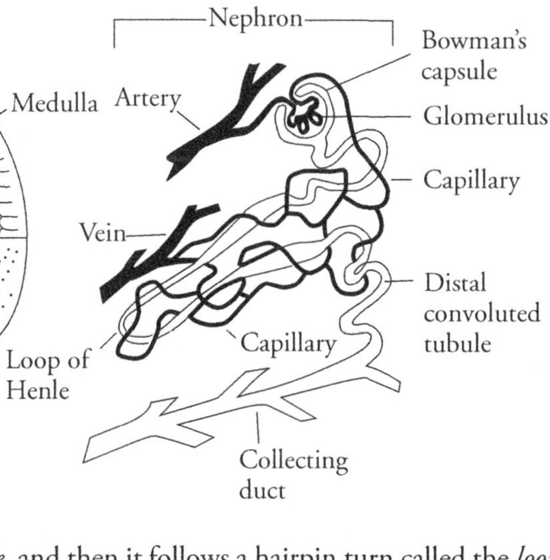

form the *proximal tubule,* and then it follows a hairpin turn called the *loop of Henle.* After the loop of Henle, the tubule coils once more as the *distal tubule.* It then enters a *collecting duct,* which also receives urine from other distal tubules.

Within the Bowman's capsule is a coiled ball of capillaries known as a **glomerulus.** Blood from the renal artery enters the glomerulus. The force of the blood pressure induces plasma to pass through the walls of the glomerulus, pass through the walls of the Bowman's capsule, and flow into the proximal tubule. Red blood cells and large proteins remain in the blood.

After plasma enters the proximal tubule, it passes through the coils, where usable materials and water are reclaimed. Salts, glucose, amino acids, and other useful compounds flow back through tubular cells into the blood by active transport. Osmosis and the activity of hormones assist the movement. The blood fluid then flows through the loop of Henle into the distal tubule. Once more, salts, water, and other useful materials flow back into the bloodstream. Homeostasis is achieved by this process: A selected amount of hydrogen, ammonium, sodium, chloride, and other ions maintain the delicate salt balance in the body.

The fluid moving from the distal tubules into the collecting duct contains urine. The urine flows through the ureters toward the urinary bladder. When the bladder is full, the urine flows through the **urethra** to the exterior.

Control of kidney function

Kidneys work to maintain homeostasis in the body. **Osmoregulation** includes processes that control solute concentrations and balance water gain and loss. The nephrons (the functional unit of the kidney) respond to the environment (via hormones) in order to keep blood osmolarity as even as possible. This osmoregulation is achieved through **negative feedback** regulation. For example, if a person drinks a large amount of water, the blood osmolarity falls below a set point. To compensate for this dilution of the blood, the kidneys increase the amount of water excreted in the urine. Conversely, if a person sweats profusely or ingests a large amount of salt, the blood osmolarity increases, thus triggering the kidneys to reabsorb water. Urine becomes more concentrated and blood osmolarity falls back down toward the set point.

Humans produce a hormone called *antidiuretic hormone* (*ADH*), also known as *vasopressin,* which is secreted by the posterior lobe of the **pituitary gland.** It regulates the osmolarity of blood by controlling the rate of water absorption in the nephron tubules. The ADH binds to receptors on the cells lining the collecting duct and causes a temporary increase in the number of aquaporin protein channels in the plasma membranes. **Aquaporin** proteins form water channels that facilitate the passive movement of water across a cell membrane; an increase in the number of aquaporins causes an increase in water reabsorption from the collecting duct back into the bloodstream.

Some individuals suffer from a condition in which they secrete very low levels of ADH. The result is excessive urination and a disease called *diabetes insipidus.* Another unrelated form of diabetes, *diabetes mellitus,* is more widespread. People with this disease produce insufficient levels of insulin. Insulin normally transports glucose molecules into the cells. But when insulin is not available, the glucose remains in the bloodstream. The glucose is removed from the bloodstream in the nephron; to dilute the glucose, the nephron removes large amounts of water from the blood. Thus, the urine tends to be plentiful.

Hormones from the **adrenal cortex** also control the content of urine. These hormones promote reabsorption of sodium and chloride ions in the tubules. Thus, they affect the water balance in the body because water flows in the direction of high sodium and chloride content.

Chapter Check-Out

Questions

1. Flatworms contain a series of excretory cells called ______.
   a. nephridia cells
   b. flame cells
   c. Malpighian cells
   d. nephrons
2. Blood enters the human kidney through the ______ and leaves through the ______.
   a. renal arteries; renal veins
   b. renal veins; renal arteries
   c. renal arteries; ureter
   d. ureter; renal veins
3. Capillaries within the Bowman's capsule make up the ______.
   a. nephron
   b. ureter
   c. distal tube
   d. glomerulus
4. Antidiuretic hormone (ADH) regulates the rate of which of the following processes?
   a. water excretion
   b. water absorption
   c. insulin absorption
   d. insulin excretion

Answers

1. b
2. a
3. d
4. b

Chapter 27

SUPPORT AND MOVEMENT

Chapter Check-In

- ☐ Comparing skeletal systems
- ☐ Studying movement in animal cells
- ☐ Contracting your muscles

Movement is one of the essential features of living things. Cellular movement is observed in one-celled amoebas, ciliates, and flagellates. Flagella whip about to produce cellular motion, while cilia beat synchronously to propel a cell.

In animals, movement is essential for locating food, escaping predators, and seeking mates. In many animals, the movement process is centered in the muscle cell, which contracts and relaxes. The contraction yields great force, which is applied against a surface by means of a skeleton.

Skeletons in Animals

Skeletal systems provide structure and protection for a variety of organisms. A water-based skeleton provides the structure necessary for movement in worms. The hard external skeleton (exoskeleton) not only provides a protective mechanism for many organisms but also assists in movement of insects. The internal skeleton (endoskeleton) present in many animals provides the structural network for support, protection, and movement.

Hydrostatic skeleton

Many animals have a water-based skeleton, or **hydrostatic skeleton.** Hydrostatic skeletons do not contain hard structures, such as bone, for muscles to pull against. Rather, the muscles surround a fluid-filled body cavity. In a worm, for example, movement occurs when muscle cells contract, and the contractions squeeze internal fluid (the hydrostatic skeleton) against the skin, causing the worm to stiffen and the body to shorten and widen. The squirming motion of a worm also depends on a hydrostatic skeleton.

Exoskeleton

A second type of skeleton, the **exoskeleton,** exists in arthropods and mollusks. In mollusks, the exoskeleton is a hard, protective outer covering. An example is a clamshell; when a clam's muscles contract, they close the shell rapidly, creating a spurt of water that propels the clam. In arthropods, the exoskeleton also provides protection and movement. Usually, wings are attached by muscles in the hard body surface, which provides a foundation for the muscle contractions. Muscle contractions raise and lower the wings, allowing flight.

Endoskeleton

Vertebrates have an internal skeleton called the **endoskeleton,** a framework of bones and cartilage (see Figure 27-1) that serves as a point of attachment for muscle. The endoskeleton thus transmits the force of muscle contractions. The endoskeleton also provides support for the body (for example, the legs) and protection (the skull).

Bone contains concentric rings of tissue in which bone cells called *osteoblasts* produce the inorganic materials (fibers and matrix) of bone. Much of this material is calcium phosphate, formed from calcium and phosphorus delivered by the blood. Living, mature bone cells called *osteocytes* are also located in the bone. Bone-destroying cells called *osteoclasts* break down bone, thus providing a turnover of bone material needed in other areas. The combination of bone cells and bone tissue comprises a unit called a *Haversian system.* Blood vessels and nerves also exist within the Haversian system.

Bones come together to form a joint, which may be immovable, such as in the **sutures** of the skull, or movable, such as in the joints of the elbow and shoulder. In a movable joint, a capsule of synovial fluid provides lubrication. Tough, fibrous tissues, known as **ligaments,** link bones to one another. Connective tissues, called **tendons,** attach muscles to bones.

Figure 27-1 The human skeleton showing the major bones of the body.

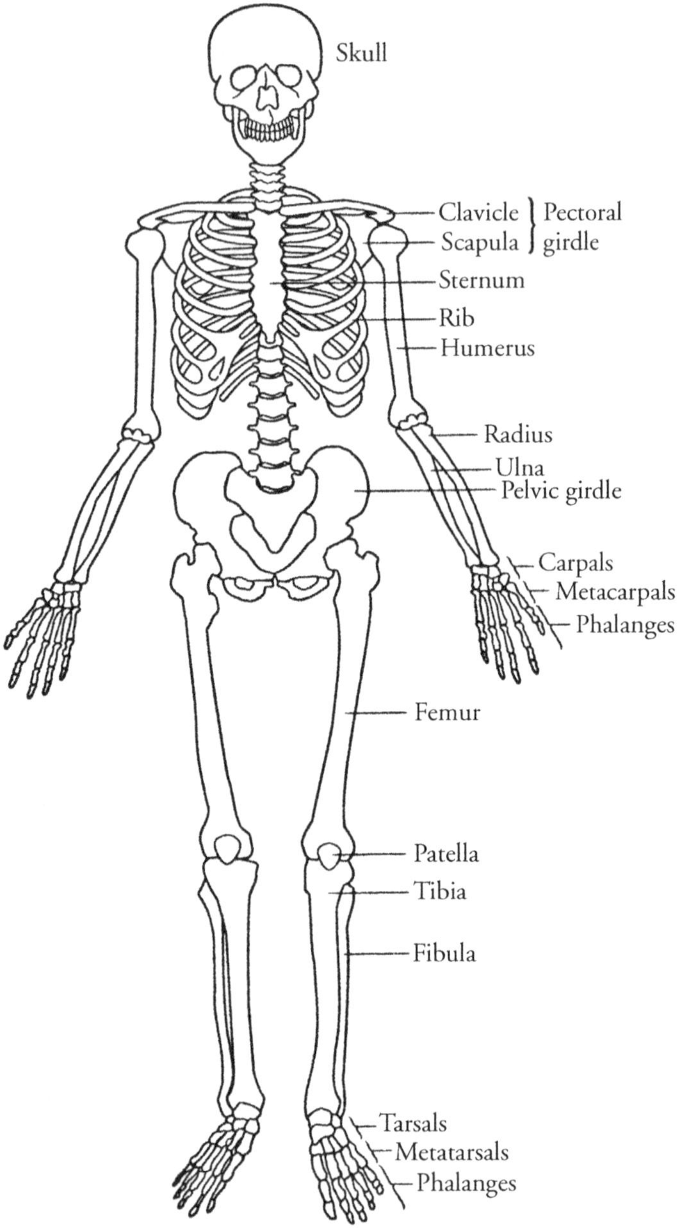

Human Muscle

Muscle is made up of thousands of muscle fibers, each composed of a single muscle cell. As shown in Figure 27-2, a muscle cell contains a series of ultramicroscopic filaments called **myofibrils.** Each myofibril is a muscle cell that contains units called **sarcomeres.** Sarcomeres contain thick microfilaments composed of the protein **myosin.** Sarcomeres also contain thin microfilaments composed of the protein **actin.** The actin and myosin filaments are arranged parallel to one another, with the myosin filaments' molecular "heads" protruding toward the actin filaments. In skeletal muscle, the overlapping actin and myosin filaments give the muscle fiber a banded, or striated, appearance. Hence, the muscle is **striated muscle.**

Figure 27-2 Anatomical structure of the muscle.

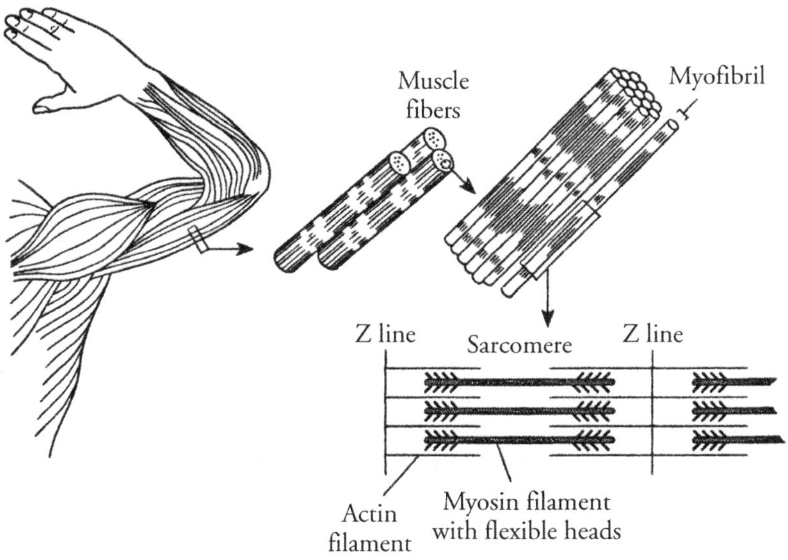

Muscle contraction

When a nerve impulse arrives at the muscle cells, it passes across the neuromuscular junction and enters the muscle cell membrane, which is known as the **sarcolemma.** The impulse spreads across the muscle cell and enters its cytoplasm, which is called *sarcoplasm.* The nerve impulse causes the actin filaments to slide across the surface of the myosin filaments. The sliding filaments pull together the ends of the muscle cell, thereby causing it to contract. The sliding filaments require that calcium ions and energy in the form of ATP be available. Two proteins called *tropomyosin* and *troponin* also function in the contraction. Cross-bridges hold the filaments together as the muscle contracts.

After the **muscle contraction** has taken place, the energy to sustain the contraction is used up, and the cross-bridges break. The filaments then slide back to their original position, and the muscle cell relaxes. There is no partial contraction of the muscle cell. Contraction is an all-or-none phenomenon.

Energy for contraction

Adenosine triphosphate (ATP) supplies the energy for muscle contraction. The reactions of glycolysis, the Krebs cycle, and the electron transport system normally produce ATP during cellular respiration (see Chapter 6). During normal activity, ATP is regenerated as it is used up during muscle contraction. When a person engages in strenuous activity, however, ATP is quickly used up and creatin phosphate is used for energy. Creatin phosphate transfers its energy to new ATP molecules, which then function as additional energy sources.

When creatin phosphate is used up, muscle cells obtain their energy solely from the process of glycolysis. Because no oxygen is available, the metabolism is anaerobic. Under these conditions, two molecules of ATP are obtained per molecule of glucose metabolized. The pyruvic acid that forms is converted to lactic acid in the muscle. Lactic acid prevents overexertion of the muscle because as lactic acid accumulates, the person experiences fatigue. The fatigue induces the person to stop exerting the muscles and breathe deeply. This breathing provides a plentiful supply of oxygen to satisfy the oxygen debt. Lactic acid is converted back to pyruvic acid, which is then metabolized through the Krebs cycle and electron transport system to provide a new supply of ATP and creatin phosphate.

Types of muscle

The human body has three major types of muscle. The muscle type discussed earlier in this chapter is *striated muscle* because the fibers' overlapping actin and myosin filaments give it a banded appearance (see Figure 27-2). This muscle is found in the limbs and is also called **skeletal muscle.** It operates under voluntary control and so is additionally known as **voluntary muscle.**

The second muscle type is **smooth muscle,** which has few actin and myosin filaments; therefore, it has few striations. Smooth muscle is found in the linings of the blood vessels, along the gastrointestinal tract, in the respiratory tract, and in the urinary bladder. Because it operates without voluntary control, it is sometimes called **involuntary muscle.**

The third muscle type is **cardiac muscle,** which is found in the heart. It has striations because it has multiple actin and myosin filaments, but it is an involuntary muscle. The actin and myosin filaments in cardiac muscle exist as intertwined branches that form a conducting network for nerve impulses.

Chapter Check-Out

Questions

1. Mature bone cells are called ______ and bone-destroying cells are called ______.
   a. osteocytes, osteoclasts
   b. osteoclasts, osteocytes
   c. osteoblasts, osteoclasts
   d. osteoclasts, osteoblasts
2. In joints, bones are linked to each other by fibrous tissues called ______.
   a. tendons
   b. muscles
   c. sutures
   d. ligaments
3. Energy for muscle contraction is supplied by ______.
   a. creatin phosphate
   b. lactic acid
   c. ATP
   d. tropomyosin
4. Skeletal muscle is ______, whereas smooth muscle is ______.
   a. involuntary, voluntary
   b. voluntary, striated
   c. involuntary, striated
   d. voluntary, involuntary

Answers

1. a
2. d
3. c
4. d

Chapter 28
CHEMICAL COORDINATION

Chapter Check-In

- ❑ Coordinating hormone secretion
- ❑ Mapping the human endocrine system

The animal body has two levels of coordination: nervous coordination (see Chapter 29) and chemical coordination. Chemical coordination is centered in a system of glands known as **endocrine glands.** These glands are situated throughout the animal body and include such organs as the pancreas, thyroid gland, and adrenal gland. The glands secrete **hormones,** a series of chemical substances composed of protein or sterol lipids.

Hormones bring about changes that help coordinate body systems in a general way. For example, the pancreas secretes insulin, which facilitates the passage of glucose into all body cells for use in energy metabolism. Another example is thyroxine, a thyroid gland secretion that regulates overall body metabolism. In contrast to chemical coordination, the nervous system coordinates functions in the animal body on a more localized level as it delivers nerve impulses to contract body muscles or regulate gland activities.

The endocrine glands secrete their hormones into the bloodstream, where the blood carries the hormones to the target organs. Because the endocrine glands have no ducts, they are often called **ductless glands.** Other glands of the body (such as the enzyme-secreting salivary glands) deliver their enzymes via ducts and are referred to as **exocrine glands.**

The structure and physiology of hormones and endocrine glands are relatively similar in all animals; the emphasis in this chapter is on the human endocrine system.

Human Endocrine System

The human endocrine system modulates several processes of the body by the function of hormones. The endocrine system secretes hormones that control how bodily functions work. Thus, the human endocrine system watches over and coordinates all the systems of the body with the use of hormones.

Pituitary gland

The **pituitary gland** is located at the base of the human brain. The gland consists of two parts: the anterior lobe (adenohypophysis) and the posterior lobe (neurohypophysis).

The anterior lobe secretes at least seven hormones. One hormone, the human growth hormone (HGH), promotes body growth by accelerating protein synthesis. This hormone is also known as *somatotropin.* A deficiency of the hormone results in dwarfism; an oversecretion results in gigantism.

Another hormone of the anterior pituitary is prolactin, also called lactogenic hormone (LH). This hormone promotes breast development and milk secretion in females. A third hormone is thyroid-stimulating hormone (TSH). The function of TSH is to control secretions of hormones from the thyroid gland. A fourth hormone is adrenocorticotropic hormone (ACTH). This hormone controls the secretion of hormones from the adrenal glands.

There are three more hormones produced in the anterior lobe of the pituitary gland. The first is follicle-stimulating hormone (FSH). In females, FSH stimulates the development of a follicle, which contains the egg cell; in males, the hormone stimulates sperm production. The next hormone is luteinizing hormone (LH). In females, LH completes the maturation of the follicle and stimulates the formation of the corpus luteum, which temporarily secretes female hormones. In males, LH is interstitial cell-stimulating hormone (ICSH), which stimulates the production of male hormones in the testes. The final hormone is melanocyte-stimulating hormone (MSH), which stimulates production of the pigment melanin.

The posterior pituitary gland stores and then releases two hormones that are produced in the hypothalamus of the brain. The first hormone is antidiuretic hormone (ADH). This hormone stimulates water reabsorption in the kidneys. It is also called *vasopressin.* The second hormone is oxytocin, which stimulates contractions in the muscles of the uterus during birth.

Thyroid gland

The **thyroid gland** lies against the pharynx at the base of the neck. It consists of two lateral lobes connected by an isthmus. The gland produces thyroxine, a hormone that regulates the rate of metabolism in the body. It also produces a second hormone, calcitonin, which regulates the level of calcium in the blood.

Thyroxine production depends on the availability of iodine. A deficiency of iodine causes thyroid gland enlargement known as a *goiter.* An undersecretion of thyroxine during childhood development results in a condition known as *cretinism* (dwarfism with abnormal body proportions and possible mental retardation). In adults, an undersecretion can also result in *myxedema* (physical and mental sluggishness). Thyroxine oversecretion results in a high metabolic rate and Graves' disease.

Parathyroid glands

The **parathyroid glands** are located on the posterior surfaces of the thyroid gland. They are tiny masses of glandular tissue that produce parathyroid hormone, also called *parathormone.* Parathyroid hormone regulates calcium metabolism in the body by increasing calcium reabsorption in the kidneys and by increasing the uptake of calcium from the digestive system.

Adrenal glands

The **adrenal glands** are two pyramid-shaped glands lying atop the kidneys. The adrenal glands consist of an outer portion, the **adrenal cortex,** and an inner portion, the **medulla.**

The adrenal cortex secretes a family of steroids called **corticosteroids.** The two main types of steroid hormones are mineralocorticoids and glucocorticoids. Mineralocorticoids such as aldosterone control mineral metabolism in the body. They accelerate mineral reabsorption in the kidneys. Mineralocorticoid secretion is regulated by ACTH from the pituitary gland. Glucocorticoids such as cortisol and cortisone control glucose metabolism and protein synthesis in the body. Glucocorticoids are also anti-inflammatory agents.

The adrenal medulla produces two hormones: **epinephrine** (adrenaline) and **norepinephrine** (noradrenaline). Epinephrine increases heart rate, blood pressure, and the blood supply to skeletal muscles. Epinephrine functions in stressful situations to promote the fight-or-flight response. Norepinephrine intensifies the effects of epinephrine. Both hormones prolong and intensify the effects of the sympathetic nervous system.

Pancreas

The pancreas is located just behind the stomach. Its endocrine portion consists of cell clusters called the **islets of Langerhans.**

The pancreas produces two hormones: insulin and glucagon. Insulin is a protein that promotes the passage of glucose molecules into the body cells and regulates glucose metabolism. In the absence of insulin, glucose is removed from the blood and excreted in the kidney, a condition called *diabetes mellitus.* Diabetes mellitus is characterized by glucose in the urine, heavy urination, excessive thirst, and a generally sluggish body metabolism.

The second pancreatic hormone, **glucagon,** stimulates the breakdown of glycogen to glucose in the liver. It also releases fat from the adipose tissue so the fat can be used for the production of carbohydrates.

Other endocrine glands

Among the other endocrine glands are the ovaries and testes. The **ovaries** secrete **estrogens,** which encourage the development of secondary female characteristics. The **testes** secrete **androgens,** which promote secondary male characteristics. Testosterone is an important androgen.

The **pineal gland** is a tiny gland in the midbrain. Its functions are largely unknown, but it seems to regulate mating behaviors and day-night cycles. The **thymus gland** is located in the neck tissues. It secretes **thymosins,** which influence the development of the T-lymphocytes of the immune system.

Prostaglandins are hormones secreted by various tissue cells. These hormones produce their effects on smooth muscles, on various glands, and in reproductive physiology. **Erythropoietin** is a hormone produced by the kidney cells. Erythropoietin functions in the production of red blood cells. **Gastrin** and **secretin** are hormones produced by digestive glands to influence digestive processes.

Chapter Check-Out

Questions

1. The anterior pituitary gland produces hormones such as ________, whereas the posterior pituitary gland produces hormones such as ______.
   - **a.** prolactin; calcitonin
   - **b.** vasopressin; calcitonin
   - **c.** prolactin; vasopressin
   - **d.** vasopressin; prolactin
2. The hormone produced in the adrenal medulla that increases the heart rate, blood pressure, and blood supply to skeletal muscles is called ______.
   - **a.** norepinephrine
   - **b.** epinephrine
   - **c.** calcitonin
   - **d.** vasopressin
3. ______ promotes the passage of glucose into cells, whereas ______ induces the breakdown of glycogen.
   - **a.** Thyroxine; insulin
   - **b.** Oxytocin; insulin
   - **c.** Insulin; oxytocin
   - **d.** Insulin; glucagon
4. ______ promotes breast development and milk secretion in females.
   - **a.** Luteinizing hormone
   - **b.** Lactogenic hormone
   - **c.** Epinephrine
   - **d.** Estrogen

Answers

1. c
2. b
3. d
4. b

Chapter 29
NERVOUS COORDINATION

Chapter Check-In

- ❑ Coordinating the human nervous system
- ❑ Sensing the external environment

Nervous coordination enables an organism's rapid response to an external or internal stimulus. Characteristic of animals only, nervous coordination is the function of the nervous system. The receptors for nervous coordination are generally located in the sense organs at the body surface, while the response in nervous coordination generally involves a gland or muscle. The function of coordination is accomplished by means of a set of signals conducted along a series of nerve cells.

Animal Nerve Cells

Animal nerve cells are specialized cells called neurons. Depending on function, these cells can be divided into sensory neurons, interneurons, and motor neurons. These three types of nerve cells coordinate with each other to receive external stimuli and to transmit the impulse to muscles or glands of the body for an appropriate response to the stimulus.

Neuron

The **neuron** is the nerve cell. Approximately 12 billion neurons exist in the human body, the great majority of them in the brain and spinal cord. The main portion of the neuron is the **cell body.** Protruding from the cell body are one or more short extensions called **dendrites** and one long extension called the **axon.** Axons are covered by a fatty layer of material called the **myelin sheath.** Bundles of axons bound together are referred to as a **nerve.**

There are three types of neurons in animals: sensory neurons, interneurons, and motor neurons. **Sensory neurons** receive stimuli from the external environment; **interneurons** (or association neurons) connect sensory and motor neurons and carry stimuli in the brain and spinal cord; **motor neurons** transmit impulses from the brain and spinal cord to the muscle or gland that will respond to the stimulus. The neurons are supported, protected, and nourished by cells of the nervous system known as **glial cells.** Together with extracellular tissue, the glial cells make up the **neuroglia.**

Nerve impulse

A nerve impulse is an electrochemical event that occurs within the neuron. In an inactive neuron, the cytoplasm is negatively charged with respect to the outside of the cell. This difference in electrical charge is maintained by the active transport of sodium ions out of the cytoplasm. A cell in this state is said to have a **resting potential,** and it is polarized.

A nerve impulse is generated when the difference in electrical charge disappears. This occurs when a stimulus contacts the tip of a dendrite and increases the permeability of the cell membrane to sodium ions. The ions rush back into the cytoplasm, and the difference in electrical charges disappears. This creates a pulse of electrochemical activity called the **nerve impulse.** A neuron displaying a nerve impulse is said to have **action potential.** The cell is depolarized.

More specifically, the influx of sodium ions into the neuron cytoplasm activates the adjacent portion of the cell membrane to admit sodium ions also. Successively, the adjacent areas of the neuron lose their differences in electrical charge, and a wave of depolarization is generated in the neuron. This wave of depolarization is the nerve impulse. After the wave of depolarization has passed, the neuron reestablishes the difference in charges by pumping potassium ions out of the cytoplasm and then pumping sodium ions in.

A neuron's action potential is an example of **positive feedback.** Unlike its more common counterpart, negative feedback, positive feedback strives to amplify a stimulus, leading to a greater response. For example, depolarization opens sodium channels, resulting in a flow of $Na+$ into the neuron. This results in further depolarization, causing more sodium channels to open.

Synapse

The nerve impulse passes down the dendrite, through the cell body, and down the axon. At the end of the axon, the impulse encounters a fluid-filled space separating the end of the axon from the dendrite of the next neuron or from a muscle cell. This space is the **synapse.** A synapse located at the junction of a neuron and muscle fiber is a neuromuscular junction.

As the impulse reaches the end of the axon, it induces changes in the cell membrane and the release of chemical substances called **neurotransmitters** (for example, acetylcholine). Molecules of neurotransmitters accumulate in the synapse and increase the membrane permeability of the next dendrite. This causes an influx of sodium ions, and a new nerve impulse is generated. After the nerve impulse has swept down the next dendrite, the neurotransmitters in the synapse are destroyed.

Reflex arc

The **reflex arc,** the simplest unit of nervous activity, involves the detection of a stimulus in the environment by sensory nerve endings, followed by impulses that travel via the sensory neurons to the spinal cord. Here, the impulses synapse with interneurons, and the interneurons generate impulses to respond to the stimulus. The impulses travel along the motor neurons to muscles or glands that respond appropriately.

In some cases, a reflex arc involves an interpretation. For this activity, interneurons transmit impulses up the spinal cord to the conscious area of the brain, where an analysis occurs.

Human Central Nervous System

The human nervous system can be conveniently subdivided into two divisions: the **central nervous system** (the brain and spinal cord) and the peripheral nervous system (the nerves extending to and from the central nervous system).

Spinal cord

The **spinal cord** of the central nervous system is a white cord of tissue passing through the bony tunnel made by the vertebrae. The spinal cord extends from the base of the brain to the bottom of the backbone. Three membranes called *meninges* surround the spinal cord and protect it. The outer tissue of the spinal cord is white (white matter), while the inner tissue is gray (gray matter).

Thirty-one pairs of projections called **nerve roots** extend out along each side of the spinal cord. The nerve roots are sites of axons belonging to sensory and motor neurons. A central canal in the spinal cord carries cerebrospinal fluid, which provides for the nutrition and gaseous needs of the cord tissue. The neurons of the spinal cord serve as a coordinating center for the reflex arc and a connecting system between the peripheral nervous system and the brain.

Brain

The brain is the organizing and processing center of the central nervous system. It is the site of consciousness, sensation, memory, and intelligence. The brain receives impulses from the spinal cord and from 12 pairs of cranial nerves coming from and extending to the senses and to other organs. In addition, the brain initiates activities without environmental stimuli.

Two major hemispheres, the left and the right hemispheres, make up the tissue of the brain. The outer portion of the brain consists of gray matter, while the inner portion is white matter. Three major portions of the brain are recognized: the hindbrain, the midbrain, and the forebrain.

The **hindbrain** consists of the medulla, cerebellum, and pons. The **medulla** is the swelling at the tip of the brain that serves as the passageway for nerves extending to and from the brain. The **cerebellum** lies adjacent to the medulla and serves as a coordinating center for motor activity; that is, it coordinates muscle contractions. The **pons** is the swelling between the medulla and the midbrain. The pons acts as a bridge between various portions of the brain.

The **midbrain** lies between the hindbrain and the forebrain. It consists of a collection of crossing nerve tracts and is the site of the *reticular formation,* a group of fibers that arouse the forebrain when something unusual happens.

The **forebrain** consists of the cerebrum, the thalamus, the hypothalamus, and the limbic system. The **cerebrum** contains creases and furrows called convolutions that permit the cerebral hemisphere to accommodate more than 10 billion cells. Each hemisphere of the cerebrum has four lobes, and activities such as speech, vision, movement, hearing, and smell occur in these lobes. Higher mental activities such as learning, memory, logic, creativity, and emotion also occur in the cerebrum.

The **thalamus** serves as an integration point for sensory impulses, while the **hypothalamus** synthesizes hormones for storage in the pituitary gland. The hypothalamus also appears to be a control center for such visceral functions as hunger, thirst, body temperature, and blood pressure. The **limbic system** is a collection of structures that ring the edge of the brain and function as centers of emotion.

Human Peripheral Nervous System

The **peripheral nervous system** is a collection of nerves that connect the brain and spinal cord to other parts of the body and the external environment. It is subdivided into the sensory somatic system and the autonomic nervous system.

The **sensory somatic system** carries impulses from the external environment and the senses. It consists of 12 pairs of cranial nerves and 31 pairs of spinal nerves. The sensory somatic system permits humans to be aware of the outside environment and react to it voluntarily.

The **autonomic nervous system** works on an involuntary basis. It consists of two groups of motor neurons and a set of knotlike groups of cell bodies called **ganglia.** Motor neurons extend to and from the ganglia to the body organs. One subdivision of the autonomic nervous system is the **sympathetic nervous system.** Impulses propagated in this system prepare the body for an emergency. They cause the heart rate to increase, the arteries to constrict, the pupils to dilate, and other changes to take place. The other subdivision is the **parasympathetic nervous system.** Impulses in this system return the body to normal after an emergency has occurred.

Human Senses

The sense organs connect the nervous system to the external environment. They are the sources of stimuli that cause a response in the nervous system, and they are the sources of all information to the human body.

Eye

In the human eye, the nerve cells are located in a single layer called the **retina,** located along the back wall of the eye. Light rays enter the eye through a curved, transparent structure called the *cornea,* and then pass through the pupil, an opening in the eyeball. The iris regulates the size of the pupil. Next, the **lens** focuses the light on the retina, which contains two types of light-sensitive cells, **rod cells** and **cone cells.** Cone cells, which detect color, are concentrated in the central portion of the retina, while rod cells, which permit vision in dim light, are concentrated at the edge of the retina. A light-sensitive pigment called **rhodopsin** functions in the detection of light.

From the eye, a series of impulses is generated for transmission to the brain. The **optic nerve** carries these impulses. The region of keenest vision, the *fovea,* is located at the center of the retina. When vision is poor, light is not focusing on the fovea, and corrective lenses are prescribed.

Ear

The ear is the organ of hearing in humans. The outer ear funnels vibrations to the **eardrum,** or tympanic membrane, which transmits the vibrations to three inner ear bones: the *malleus* (hammer), the *incus* (anvil), and the *stapes* (stirrup). These bones transmit the vibrations to the inner ear, where the receptor of hearing, the cochlea, is located.

The **cochlea** is a snail-like series of coiled tubes within the skull. As the ear bones vibrate, they push and pull a membrane at one end of the cochlea, causing fluid within the tubules to vibrate. The vibrations are detected by sensitive hair cells, and nerve impulses are generated. The **auditory nerve** carries the impulses to the brain for interpretation.

Taste and smell

Specialized receptor cells called **chemoreceptors** transmit taste and smell. Chemoreceptors of the human tongue distinguish four different tastes: sweet, sour, salty, and bitter. In the human nose, chemoreceptors detect a variety of scents, including minty, floral, musky, putrid, and pungent.

In both taste and smell, chemoreceptors are stimulated by molecules and ions that reach the tongue and nose. Liquid materials affect the chemoreceptors in the taste buds of the tongue, while gaseous molecules affect the chemoreceptors in the upper reaches of the nose. The **olfactory nerve** carries nerve impulses from the nose to the brain for interpretation.

Other senses

The other senses of the body include receptors for touch, pain, temperature, and balance. Touch and pain receptors, called **Pacinian corpuscles,** are located in the skin, muscles, and tendons. The sense of balance is centered in the semicircular canals of the inner ear. Visceral senses include stretch receptors in the muscles, as well as carbon dioxide receptors in the arteries.

Chapter Check-Out

Questions

1. ______ receive stimuli from the external environment, whereas ______ transmit impulses from the brain to muscles or glands.
   a. Motor neurons; sensory neurons
   b. Sensory neurons; interneurons
   c. Sensory neurons; motor neurons
2. ______ provide(s) for the nutrition and gaseous needs of the cord tissue.
   a. Cerebrospinal fluid
   b. Nerve roots
   c. The reflex arc
3. The ______ consists of the cerebrum, the thalamus, the hypothalamus, and the limbic system.
   a. hindbrain
   b. forebrain
   c. midbrain
4. The ______ nervous system permits humans to voluntarily react to the outside environment, whereas the ______ nervous system functions on an involuntary basis.
   a. parasympathetic; sensory somatic
   b. peripheral; sympathetic
   c. sensory somatic; autonomic

Answers

1. c
2. a
3. b
4. c

Chapter 30
HUMAN REPRODUCTION

Chapter Check-In

- ❑ Understanding the male and female reproductive systems
- ❑ Fertilizing an egg
- ❑ Developing a fetus

Reproduction is an essential process for the survival of a species. The functions of the reproductive systems are to produce reproductive cells, the **gametes,** and to prepare the gametes for fertilization. In addition, the male reproductive system delivers the gametes to the female reproductive tract. The female reproductive organs nourish the fertilized egg cell and provide an environment for its development into an embryo, a fetus, and a baby.

Human reproduction takes place by the coordination of the male and female reproductive systems. In humans, both males and females have evolved specialized organs and tissues that produce **haploid** cells, the sperm and the egg. These cells fuse to form a zygote that eventually develops into a growing fetus. A hormonal network is secreted that controls both the male and female reproductive systems and assists in the growth and development of the fetus and the birthing process.

Male Reproductive System

The male reproductive organs are the **testes** (or testicles). The testes are two egg-shaped organs located in a pouch called the **scrotum** outside the body. In the scrotum, the temperature is a few degrees cooler than body temperature. The testes develop in the abdominal cavity before birth and then descend into the scrotum.

Sperm production in the testes takes place within coiled passageways called **seminiferous tubules.** Within the walls of these tubules, primitive cells called **spermatogonia** undergo a series of changes and proceed through meiosis to yield **sperm cells.** Each human sperm cell has 23 chromosomes.

The sperm cells mature in a tube called the **epididymis,** which is located along the surface of the testes. The hormone that stimulates sperm cell production is follicle-stimulating hormone (FSH). A second hormone important in reproduction is interstitial cell-stimulating hormone (ICSH), which acts on interstitial cells located between the seminiferous tubules. The interstitial cells secrete male hormones, including testosterone. The male hormones regulate the development of secondary male characteristics.

The organ responsible for carrying the sperm cells to the female is the penis. Within the penis, the sperm cells are carried in a tube, the **urethra.** During periods of sexual arousal, the penis becomes erect as blood fills its spongelike tissues. The sperm cells are mixed with secretions from the prostate gland, seminal vesicles, and Cowper's glands. These secretions and the sperm cells constitute the **semen.**

Female Reproductive System

The organs of female reproduction include the **ovaries,** two oval organs lying within the pelvic cavity, and adjacent to them, two **fallopian tubes.** Also known as **oviducts,** the fallopian tubes are the passageways that egg cells enter after release from the ovaries. The fallopian tubes lead to the **uterus** (womb), a muscular organ in the pelvic cavity. The inner lining of the uterus, called the *endometrium,* thickens with blood and tissue in anticipation of a fertilized egg cell. If fertilization fails to occur, the endometrium degenerates and is shed in the process of menstruation.

The opening at the lower end of the uterus is a constricted area called the **cervix.** The tube leading from the cervix to the exterior is a muscular organ called the **vagina.** During periods of sexual arousal, the vagina receives the penis and the semen. The sperm cells in the semen pass through the cervix and uterus into the fallopian tubes, where fertilization takes place.

In the human female, egg cell production begins before birth, when about 2 million primitive cells known as **oogonia** accumulate in the ovaries. These oogonia are formed in the early stages of meiosis. After the age of puberty, the oogonia develop into primary **oocytes** and then into **egg**

cells at a rate of one per month. Egg cell production occurs by the process of meiosis.

The egg cells develop within the ovary in a cluster of cells called the **Graafian follicle,** which secretes the female hormone **estrogen** that regulates the development of secondary female characteristics. Egg cell development within the follicle requires approximately 14 days. The development is controlled by two hormones: follicle-stimulating hormone (FSH) and luteinizing hormone (LH). Both hormones are secreted by the anterior lobe of the pituitary gland.

During the 14 days of egg cell development, the endometrium increases its supply of blood and nutrients in anticipation of a fertilized egg cell. On about the 14th day, the release of the egg cell from the follicle takes place. This process is called **ovulation.** The egg cell is swept into the fallopian tube and begins to move toward the uterus. Meanwhile, the follicle is changed into a mass of cells known as the **corpus luteum.** The LH stimulates the conversion. The corpus luteum then secretes the hormone **progesterone,** which together with estrogen continues to regulate the buildup of tissue in the endometrium and inhibit contractions of the uterus.

The egg cell remains alive in the fallopian tubes for 24 to 72 hours. If fertilization by a sperm cell fails to occur, the egg cell moves toward the uterus, and the corpus luteum begins to degenerate. This degeneration causes the levels of progesterone and estrogen to drop off. Within 2 weeks, the hormone levels decline to a point where they cannot inhibit contractions of the uterus. Uterine contractions then occur, and the endometrium is released in the process of **menstruation.** Follicle development begins again, but in the opposite ovary.

Fertilization and Development

For fertilization to occur, sperm cells must be released in the vagina during the period that the egg cell is alive. The sperm cells move through the uterus into the fallopian tube, where one sperm cell may fertilize the egg cell. The fertilization brings together 23 chromosomes from the male and 23 chromosomes from the female, resulting in the formation of a **fertilized egg cell** with 46 chromosomes. The fertilized cell is a **zygote.**

The zygote undergoes mitosis to form two identical cells that remain attached. This takes place about 36 hours after fertilization. Mitosis then occurs more frequently. Soon, a solid ball of cells, a **morula,** results.

Morula formation occurs about 6 days after fertilization. During that time, the cells are moving through the fallopian tube. Within the next 2 days, a hollow ball of cells called a **blastocyst** forms. The blastocyst enters the uterus. At one end of the blastocyst, a group of cells called the **inner cell mass** continues to develop.

About 8 days after fertilization, the blastocyst implants itself in the endometrium of the uterus. During implantation, the outer cells take root in the endometrium. This outer layer of cells, called the **trophoblast,** gives rise to projections that form vessels. These vessels merge with the maternal blood vessels to form the **placenta.** The trophoblast also develops into three membranes: the amnion, the chorion, and the yolk sac membrane.

The inner cell mass undergoes changes to form three germ layers known as the ectoderm, the mesoderm, and the endoderm. The **ectoderm** becomes the skin and nervous system, the **mesoderm** becomes the muscles and other internal organs, and the **endoderm** becomes the gastrointestinal tract. The **embryo** is formed at about the fourth week, when all the organs of the body have taken shape.

Embryonic Stage

At the age of 4 weeks, the embryo is about the size of a pea. A primitive heart is beating, the head is defined with rudimentary eyes and ears, and tiny bumps represent arms and legs. The embryo also contains a primitive nervous system, and the head has begun to enlarge. A cartilage skeleton has appeared, and muscles have taken shape.

By the end of 8 weeks, the embryo is somewhat human-looking. Facial features are evident, and most of the organs are well developed. From this point onward, development consists chiefly of growth and maturation. The embryo is about 1.5 inches in length. Henceforth it is known as a **fetus.**

Nourishment of the embryo, and then the fetus, is accomplished through the placenta. The maternal and embryonic blood supplies meet at this organ, but the blood does not mix. Instead, diffusion accounts for the passage of gases, nutrients, and waste products across the membranous barriers. The placenta is also an endocrine gland because it secretes estrogen and progesterone to continue to inhibit follicle development and maintain the integrity of the endometrium. As the embryo becomes a fetus, it moves away from the placenta, and a length of tissue called the **umbilical cord** becomes its source of attachment to the maternal blood supply.

Fetal Development

During the third month, the fetus definitely resembles a human, but the head is relatively large. During the ensuing months, the remainder of the body increases in size proportionally. Cartilage is replaced by bone, and the reproductive organs develop.

During the fourth month, the length of the fetus increases to about 6 inches. The heartbeat can be heard through the mother's abdominal wall, and the fetus moves about. Distinctive movements can be felt at the fifth month, and by the sixth month the fetus weighs almost 2 pounds. By the end of 6 months, the fetus might be able to survive outside the mother's body, but it would have little fat in its skin, so temperature control would be a problem. By the end of the ninth month, the fetus has an average length of about 20 inches and a typical weight of 6 to 8 pounds.

Birth

Childbirth is a complex process regulated by various hormones. When diminished levels of progesterone remove the inhibition on uterine contractions, the uterine muscles contract. The posterior pituitary gland releases the hormone **oxytocin,** which stimulates further contractions.

During the first stage of labor, the cervix opens and the baby descends into the birth canal, or vagina. By this time, the amniotic sac has broken. In the second stage of labor, the baby passes through the birth canal, assisted by painful uterine contractions. The baby's head normally appears first, but in a breech birth the buttocks may appear first. In the third stage, the placenta (afterbirth) is delivered.

Childbirth provides another perfect example of **positive feedback.** Once oxytocin spurs uterine contractions to begin, the baby's head presses harder against the cervix. This cervical pressure signals the pituitary gland to release more oxytocin, thus further increasing contractions. This positive-feedback loop continues until the baby is born.

Chapter Check-Out

Questions

1. Sperm production takes place in the ______.
   a. scrotum
   b. seminiferous tubules
   c. prostate
2. Egg cells develop in a cluster of cells called the ______, which secretes ______.
   a. cervix; luteinizing hormone
   b. vagina; estrogen
   c. Graafian follicle; estrogen
3. The placenta is considered a(n) ______ gland because it secretes estrogen and progesterone.
   a. adrenal
   b. endocrine
   c. parathyroid
4. During development, the ______ becomes the skin and the ______ becomes the gastrointestinal tract.
   a. mesoderm; ectoderm
   b. ectoderm; endoderm
   c. endoderm; ectoderm

Answers

1. b
2. c
3. b
4. b

Chapter 31
ECOLOGY

Chapter Check-In

- ❑ Occupying a defined area
- ❑ Defining a community
- ❑ Interacting to form ecosystems

Ecology is the discipline of biology that is concerned primarily with the interaction between and among organisms and their environments. There are many levels of organization in the world of living things, and each level has features not displayed by any other. The important levels of organization are population, community, ecosystem, biosphere, and biome.

Populations

A **population** is a group of individuals belonging to one species usually occupying a defined area at a defined time. Populations of living things interact with other populations of their own kind, with populations of other species, and with physical aspects of their environment. As opposed to individual behavior, members of a population may participate in group behavior in order to increase the species' chances of survival. This includes members of the same species forming a protective group (e.g., a school of fish or a flock of birds) and engaging in cooperative behavior such as hunting or migrating.

A population's growth proceeds until reaching certain environmental limits. When a population has reached the maximum size that the environment can support indefinitely, assuming consistent conditions, the population is said to have reached its **carrying capacity** (see Figure 31-1). The growth of a population cannot exceed the carrying capacity of its environment forever. There may be an initial overshoot phase, where the numbers of individuals are too many for the environment to sustain. Likewise, the numbers of individuals may temporarily dip below the carrying capacity.

Figure 31-1 Carrying capacity and population size.

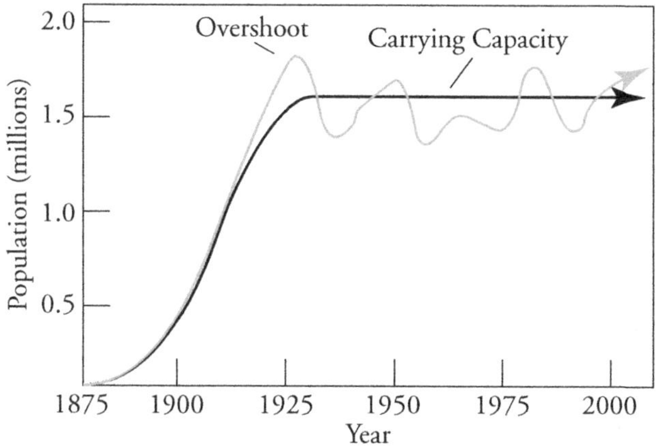

The growth of a population passes through stages. There is an initial lag period of minimal growth, followed by an exponential growth period of maximum population growth. The curbs that limit population growth may include the crowding of a population to the point that resources like food and water become scarce, an example of a density-dependent factor affecting population size. Other density-dependent factors include disease, competition, predation, and territoriality. Density-independent factors that affect population size include weather-related events.

Communities

Communities of plants, animals, and other organisms may be found in such places as a desert, a salt marsh, or a forest. A **community** is defined as multiple populations interacting with one another at the same time and in the same place. Within a community, each population of organisms has a habitat and may occupy one or more niches. The *habitat* is the physical place where the organisms live, while the *niche* is the role or roles that the population plays in the life of the community. The niche is the population's function and position in the ecosystem, and it reflects the population's relationship to other populations. A population's niche is defined by how and when it reproduces, what time of the day and year it is most active, what climatic factors it can withstand, what it eats, and so forth. The *competition exclusion principle* suggests that two species cannot occupy the same niche in the same place at the same time indefinitely.

Two populations living together in a community in a close and permanent association is called **symbiosis.** If the symbiotic relationship is mutually beneficial, it is known as **mutualism.** Lichens represent an example of mutualism. If one population receives a benefit from an association, while the other is neither benefited nor harmed, the symbiosis is **commensalism.** Humans and the bacteria of their intestines exist commensally. Another type of symbiosis is **parasitism,** in which one population benefits while the other is harmed. The microorganisms that cause human disease are considered **parasites.** A final type of symbiosis is **synergism.** In this instance, two populations accomplish together what neither population could accomplish on its own.

Populations within a community may interact; indeed, one population may capture and feed on the other. Such a relationship is known as **predation.** Predators have more than one prey species, but they normally feed upon the most abundant prey species available. As a prey population decreases, the predator switches to a more abundant species. This change causes fluctuations in population sizes.

Natural selection favors the most effective predator, while also favoring the prey that can escape predation. Among the adaptations that help prey to escape are poisonous toxins, chemical adaptations, warning coloration, camouflage, and mimicry.

In a community, the orderly and predictable replacement of populations over a given period of time in a given area is called *succession.* Primary succession occurs in lifeless areas where soil has not yet formed. An area covered by a lava flow would undergo primary succession. In secondary succession, communities are established on a site previously occupied by a population. In each succession, certain populations dominate and then decline, to be superseded by new dominant populations. A community at the last stage of succession is called the **climax community.**

Ecosystems

Interactions between communities and their abiotic physical environments form systems known as **ecosystems.** One of the major phenomena underlying an ecosystem is the flow of energy. Because photosynthesizing organisms trap the energy in an ecosystem, they are **producers.** Because certain organisms in the community meet their energy needs by feeding on these producers, they are **consumers.** Producers are autotrophs, while consumers are usually heterotrophs. Primary consumers feed directly on plants, while secondary consumers (carnivores) feed on the animals that eat the plants. The energy flow forms a one-way pattern; the energy is used for metabolic reactions and the remainder is given off as heat.

The linear transfer of food energy from producers to consumers is known as the **food chain.** Many food chains intertwine in a complex manner to form a **food web. Decomposers,** the organisms of decay, are found at each level of food chains and food webs. The organisms of decay are usually bacteria and/or fungi.

The **food pyramid** is a way of expressing the availability of food in an ecosystem at a successive number of trophic levels. The number of producers, always at the base of the pyramid, is high; the number of consumers, at the top of the pyramid, is low. The difference in numbers occurs because only a small percentage of the food energy available at one level can be passed on to the next. The total dry mass of food at each level of the pyramid is called the **biomass.**

Another phenomenon of an ecosystem is the recycling of elements. Carbon, nitrogen, and phosphorus typify those elements that are recycled. Much of the carbon is recycled in respiration, but more is recycled in decomposition, principally by bacteria and fungi.

Nitrogen, which is vital for the synthesis of proteins and nucleic acids, is released to the atmosphere as waste by bacteria. Nitrogen is brought back into the food chain by nitrogen-fixing bacteria that exist on the roots of plants called legumes (peas, beans, alfalfa, and clover). The bacteria trap the nitrogen, form ammonium ions, and make these ions available to plants for amino-acid synthesis.

Ecosystems are built on the complex interactions between different communities. These interactions help to maintain relatively consistent numbers and species of organisms, as long as conditions remain stable. Any disruption to the ecosystem (e.g., hunting, flooding, or rising sea level) may result in a new ecosystem.

Biospheres and Biomes

The **biosphere** is the blanket of living things that surrounds the substratum of Earth. The biosphere is composed of living organisms, as well as the physical environment. The physical environment includes the rocky material of Earth's crust, the water on or near Earth's surface, and the thin blanket of gas surrounding Earth. All life is confined to a 5-mile vertical space around Earth's surface.

Ecologists study the living components of the biosphere in subunits called **biomes.** A biome is a group of communities with stable ecosystems dominated by a particular climax community. Deserts, forests, and prairies are examples of biomes. Other examples are the tundra, *taiga* (the southern edge of the tundra), and temperate forests. Each biome represents a unique situation in which the ecosystem is defined by the environment. The broad diversity of living things that characterizes Earth exists in the biome. Each type of living thing is adapted to its own habitat and niche within the biome. The general composition of a biome remains uniform, but local differences arise as a result of population fluctuations, floods, fire, and other ecological factors.

Chapter Check-Out

Questions

1. The two types of Paramecium, *P. aurelia* and *P. caudatum,* are closely related species. In an experiment, each population, when grown alone under constant conditions, was able to grow and flourish. When put together, *P. aurelia* was unable to survive, yet *P. caudatum* continued to flourish. This experiment and its results are an example of which of the following?
   a. carrying capacity
   b. density-independent factors
   c. competitive exclusion
2. The term that describes two populations living together in a community and forming a long-lasting relationship is ______.
   a. mutualism
   b. symbiosis
   c. commensalism
3. Organisms that trap the energy in an ecosystem are called ______.
   a. consumers
   b. predators
   c. producers

Answers

1. c
2. b
3. c

REVIEW QUESTIONS

Use this review to practice what you've learned in this book. After you work through the review questions, you're well on your way to achieving your goal of understanding the key concepts of biology.

Questions

Chapters 1–4

1. Describe the different components that make up an atom.
2. A ______ is a precise arrangement of atoms of different elements.
   a. compound
   b. protein
   c. carbohydrate
   d. molecule
3. The ______ organelle is the site of protein and lipid processing and packaging.
   a. endoplasmic reticulum
   b. Golgi apparatus
   c. nucleus
   d. mitochondria

Match the following five types of movement with their definitions:

4. diffusion
5. osmosis
6. facilitated diffusion
7. active transport
8. endocytosis

   a. movement across the membrane from a region of lower concentration to a region of higher concentration at the expense of ATP
   b. the process by which the plasma membrane engulfs particles as they enter the cell surface
   c. movement from a region of high concentration to a region of low concentration that is assisted by proteins
   d. movement of water across a plasma membrane from a region of higher concentration to a region of lower concentration
   e. movement of molecules across a plasma membrane from a region of higher concentration to a region of lower concentration

Chapters 5–6

9. Within the carbon-fixing reactions of photosynthesis, CO_2 becomes attached to a five-carbon compound called ______.

   a. phosphoglycerate
   b. ribose
   c. ribulose bisphosphate
   d. ribulose diphosphate carboxylase

10. ______ is the process by which glucose is broken down to form two molecules of pyruvic acid.

   a. Glycolysis
   b. The Krebs cycle
   c. Chemiosmosis
   d. The Calvin cycle

11. The final electron acceptor of respiration is ______.

a. NAD

b. O_2

c. H_2O

d. FAD

12. For every one molecule of glucose, the Krebs cycle forms ______ molecules of ATP.

a. 2

b. 4

c. 6

d. 10

Chapters 7–8

13. Describe the genetic relevance of prophase I during meiosis.

Chapters 9–11

14. List the three Mendelian laws of genetics.

15. If the length of a giraffe's neck is determined at the locus *T* (tall), where tall necks are dominant over small necks, what are the possible genotypes of the parent generation if the resulting offspring contain a phenotypic ratio of 4:0?

16. What is the genotype and phenotype of the F_1 generation resulting from a cross between two pink snapdragons?

17. An enzyme called ______ adds new nucleotides to a growing DNA strand, whereas ______ creates a permanent bond between nucleotides.

a. ligase, polymerase

b. polymerase, ligase

c. polymerase, helicase

d. ligase, helicase

18. Describe the lactose operon.

Match the following three types of RNA molecules with their functions:

19. tRNA

20. mRNA

21. rRNA

- **a.** receives the genetic code in the DNA and carries the code into the cytoplasm
- **b.** carries the amino acid to the ribosomes for protein synthesis
- **c.** used to manufacture ribosomes

22. True or False: Hemophilia is a sex-linked condition that is linked to the Y chromosome.

Chapters 12–14

23. True or False: The first primitive cells were called proteinoids.

24. True or False: The endosymbiotic theory suggests that bacteria were engulfed by larger cells and eventually became the mitochondria.

Chapters 15–18

Match the following three divisions of Fungi with the organisms they contain:

25. Basidiomycetes

26. Ascomycetes

27. Zygomycetes

- **a.** mildews
- **b.** mushrooms
- **c.** bread mold

Match the following five divisions of algae with the types of algae they contain:

28. Pyrrophyta

29. Rhodophyta

30. Chlorophyta

31. Phaeophyta

32. Chrysophyta

- **a.** green algae
- **b.** kelp
- **c.** dinoflagellates
- **d.** diatoms
- **e.** red algae

Chapters 19–20

33. Describe the structural features of a flowering plant.

34. Describe the four different tissue systems of a vascular plant.

Chapters 21–22

Match the following five classes of vertebrates with the organisms they contain:

35. Agnatha

36. Amphibia

37. Osteichthyes

38. Aves

39. Placodermi

- **a.** the earliest known fishes with jaws
- **b.** fishes that have no jaws
- **c.** frogs
- **d.** birds
- **e.** bony fishes

Chapters 23–30

40. The portion of the brain that contains the higher mental activities such as learning and memory is called the ______.
    a. hindbrain
    b. midbrain
    c. cerebellum
    d. cerebrum

41. The space located at the junction of a neuron and a muscle fiber is called ______.
    a. a synapse
    b. action potential
    c. resting potential
    d. a neuromuscular junction

42. Vasopressin is produced in the ______, whereas prolactin is produced in the ______.
    a. posterior pituitary, anterior pituitary
    b. anterior pituitary, thyroid
    c. thyroid, anterior pituitary
    d. posterior pituitary, thyroid

43. Compare and contrast the hydrostatic skeleton, exoskeleton, and endoskeleton.

44. Describe the types of muscle in the human body.

45. Explain the fertilization and development process in humans.

Chapter 31

Match the following four relationships with their definitions:

46. commensalism

47. parasitism

48. synergism

49. mutualism

- **a.** a relationship where two populations accomplish together what neither population could accomplish on its own
- **b.** a relationship where one population benefits while the other population is harmed
- **c.** a relationship that is mutually beneficial
- **d.** a relationship where one population benefits while the other is neither benefited nor harmed

50. True or False: Biomes are the blanket of living things that surround the substratum of Earth.

Answers

Chapters 1–4

1. Provide your own answer (see Chapter 2).
2. d
3. b
4. e
5. d
6. c
7. a
8. b

Chapters 5–6

9. c
10. a
11. b
12. a

Chapters 7–8

13. Provide your own answer (see Chapter 8).

Chapters 9–11

14. Mendel's law of dominance; Mendel's law of segregation; Mendel's law of independent assortment (see Chapter 9).
15. *TT* × *tt*; *TT* × *TT*; *tt* × *tt*
16. *RR* : *Rr* : *rr* (1:2:1); Red, pink, white (1:2:1)
17. b
18. Provide your own answer (see Chapter 10).
19. b
20. a
21. c
22. False. Hemophilia is linked to the X chromosome.

Chapters 12–14

23. False. The first cells were called protocells.
24. True

Chapters 15–18

25. b
26. a
27. c
28. c
29. e
30. a
31. b
32. d

Chapters 19–20

33. Provide your own answer (see Chapter 19).
34. Provide your own answer (see Chapter 20).

Chapters 21–22

35. b

36. c

37. e

38. d

39. a

Chapters 23–30

40. d

41. d

42. a

43. Provide your own answer (see Chapter 27).

44. Provide your own answer (see Chapter 27).

45. Provide your own answer (see Chapter 30).

Chapter 31

46. d

47. b

48. a

49. c

50. False. The biosphere is the blanket of living things.

GLOSSARY

absorption: The process in which nutrients enter cells of the villi and then move across the cells and enter blood vessels.

acids: Compounds that release hydrogen ions (H+) when the compounds are placed in water.

actin: A protein filament within the sarcomeres of muscle cells.

action potential: Occurs when a neuron is displaying a nerve impulse.

active site: The portion of an enzyme that interacts with the substrate.

active transport: The movement of molecules across a membrane from a region of low concentration to a region of high concentration that requires the expenditure of energy (ATP).

adenosine diphosphate (ADP): A product of adenosine triphosphate (ATP) breakdown.

adenosine triphosphate (ATP): The chemical substance that serves as the currency of energy in cells.

adrenal cortex: The outer portion of the adrenal gland.

adrenal glands: Two glands lying atop the kidneys that produce a family of steroids.

aerobic: With oxygen.

agar: Solid growth medium for culturing bacteria; made from a derivative of red algae.

algae: A large number of photosynthetic organisms that are not classified as plants.

alleles: Different forms of the same gene.

alveoli: Microscopic air sacs that are surrounded by a rich network of blood vessels in mammalian lungs that function in gas exchange; the air sacs are at the end of the bronchioles.

amino acids: The building blocks of proteins.

amoeba: Single-celled organism with no distinct shape; member of the phylum Sarcodina.

anabolism: The process of synthesizing large molecules by joining smaller molecules together.

anaerobic: Without oxygen.

anaphase: A phase during mitosis in which chromatids separate to become visible chromosomes and migrate to opposite poles.

anaphase I: A phase during meiosis in which homologous chromosomes separate.

anaphase II: A phase during meiosis II in which the centromeres divide and the chromosomes separate from one another. Sister chromatids separate.

androgens: Hormones, such as testosterone, produced from the testes that promote secondary male characteristics.

Animalia: The kingdom that includes animals.

antibodies: Proteins released by B cells and released into the circulatory system that bind to inactive antigens such as bacteria and viruses.

antibody-mediated immunity: The process by which antibodies bind to antigens and destroy the microorganisms in a highly specific manner.

anticodon: The complementary codon to mRNA codon that is present on a tRNA molecule.

antigens: The immune-stimulating polysaccharides on the surfaces of cells.

aorta: The major artery of the human circulatory system that receives blood from the left ventricle.

appendix: A small fingerlike organ that may be a vestige of larger organs functional in human ancestors.

aquaporins: Channel proteins that facilitate the passive movement of water across a cell membrane.

archaea: Ancient prokaryotes that have a different ribosomal structure, membrane composition, and cell wall composition than other bacteria.

arteries: The channels through which blood flows away from the heart.

atom: The smallest part of an element that can enter into various combinations with atoms of other elements.

atrium: A thin-walled receiving chamber in the human heart in which blood accumulates.

auditory nerve: The nerve within the ear that carries impulses to the brain for interpretation.

autonomic nervous system: A subdivision of the peripheral nervous system that works on an involuntary basis. It is divided into the sympathetic and parasympathetic nervous systems.

autosomes: The 22 pairs of human chromosomes that are not sex chromosomes.

autotrophs: Certain bacteria, plants, and protists that synthesize their own food via photosynthesis.

axon: The long extension of a neuron.

bacilli: The rod-shaped bacteria (singular, *bacillus*).

bark: The structure of vascular plants formed between the phloem and the cork.

bases: Compounds that attract hydrogen atoms when placed in water.

basophils: White blood cells that function in allergic responses.

bicuspid (mitral) valve: The valve that leads into the left ventricle of the human heart.

binomial name: The scientific name of an organism, which contains two elements: genus and species.

biomass: The total dry mass of food at each level of the food pyramid.

biome: A group of communities dominated by a particular climax community, such as deserts, forests, and prairies.

biosphere: The blanket of living things that surrounds the substratum of Earth.

blastocyst: A hollow ball of cells resulting after the morula has passed through the fallopian tubes and enters the uterus.

blood clotting: The process in which platelets adhere to the walls of damaged blood vessels, setting off a series of processes leading to the formation of a patchy mesh at the injury site.

B-lymphocytes: White blood cells within the lymph nodes; stimulated by microorganisms or other foreign materials in the blood; also called B cells.

Bowman's capsule: An enlarged cup-like structure below the nephron in the human kidney.

bronchi: Two large tubes at the lower end of the trachea (singular, *bronchus*).

bronchioles: The branches formed from the bronchi.

capillaries: The microscopic blood vessels between the arteries and the veins.

carbohydrates: The primary energy source for living things; composed of carbon, hydrogen, and oxygen.

cardiac muscle: The involuntary muscle found in the human heart; contains actin and myosin filaments.

carnivores: Animals that eat other animals.

carrying capacity: The maximum population size that an environment can sustain indefinitely.

catabolism: The breakdown or digestion of large, complex molecules.

cecum: A sac that is the meeting point of the small and large intestines.

cell body: The main portion of the nerve cell.

cell cycle: Many repetitions of cellular growth and reproduction; divided into interphase, mitosis, and cytokinesis.

cell-mediated immunity: The process in which the T-lymphocytes interact with the microorganisms cell-to-cell and destroy them.

cells: The fundamental units of living things.

cellular respiration: The process by which animals and other organisms obtain the energy available in carbohydrates.

cellulose: Carbohydrate used primarily as a structural polymer; structural component of plant cell walls

cell wall: A strong membrane outside the plasma membrane present in certain cells, such as bacteria and plants.

central nervous system: The portion of the nervous system that consists of the brain and the spinal cord.

centriole: A cylinderlike organelle that assists in chromosomal migration during mitosis.

centromere: The place of attachment of the sister chromatids during prophase in mitosis.

cerebellum: A portion of the hindbrain that serves as a coordinating center for motor activity.

cerebrum: The portion of the forebrain that controls higher mental activity, such as learning, memory, logic, creativity, and emotion.

cervix: The opening at the lower end of the uterus.

chemiosmosis: The subdivision of cellular respiration in which the energy given off by electrons is used to pump protons across a membrane and synthesize ATP.

chemoreceptors: The specialized receptor cells that transmit smell and taste.

chlorophyll: Green pigment that makes up a photosystem that absorbs energy from the sun during photosynthesis.

chloroplast: An organelle within green plants in which photosynthesis occurs.

chordates: Animals with notochords and a dorsal nerve cord, including reptiles, amphibians, birds, and mammals.

chromatid: One copy of a replicated chromosome.

chromatin: Compacted DNA and protein.

chromosomes: Linear units of DNA.

chyme: A soupy liquid formed in the stomach from the churning of the bolus with gastric juices.

circulatory system: The transport system in animals that moves nutrients to the tissues and removes metabolic wastes for excretion.

class: A grouping of similar orders.

climax community: A community at the last state of succession that has reached a steady state.

cloning: The ability to make a genetic replica of a cell or even an entire organism (also called *reproductive cloning*).

cocci: Spherical bacteria (singular, *coccus*).

cochlea: A snail-like series of coiled tubes within the skull that assist hearing.

codominance: An allele combination in which two characteristics are equally expressed.

coenzymes: Organic molecules that act as cofactors, such as NAD and FAD.

cofactors: Ions or molecules that associate with enzymes and are required for enzymatic reactions to take place.

commensalism: A relationship in which one population receives a benefit from an association, while the other is neither benefited nor harmed.

community: Interacting populations in the same habitat at the same time.

comparative anatomy: The comparison of anatomical structures of modern-day organisms with fossils to yield clues to the type of organisms that roamed Earth long ago.

comparative biochemistry: The comparison of biochemical processes of modern-day organisms with fossils and ancient species; modern biochemistry indicates there is a biochemical similarity in all living things.

compound: A molecule that contains atoms of more than one element.

concentration gradient: How the concentration of molecules differs from one location to another.

cone cells: The cells of the eye that detect color.

consumers: Organisms within an ecosystem that meet their energy needs by feeding on producers.

cork: A tough tissue that combines with the phloem to become the bark of vascular plants.

coronary arteries: The arteries that supply the heart muscle with blood.

corpus luteum: The mass of cells derived from the female follicle that secretes progesterone.

corticosteroids: The steroids secreted from the adrenal glands.

crossing over: A process during prophase I when homologous pairs exchange genes.

cyanobacteria: Members of the domain Bacteria that are photosynthetic and are found in the soil and in freshwater or saltwater environments. Formerly known as *blue-green algae*.

cytochromes: Molecules that accept and release electrons in an electron transport system.

cytokinesis: The process immediately following mitosis in which the cytoplasm divides into two separate cells.

cytoplasm: Semiliquid substance that composes the foundation of the cell's volume and contains the organelles.

cytoskeleton: An interconnected system of fibers, threads, and interwoven molecules that give structure to a cell.

cytosol: See *cytoplasm*.

decomposers: The organisms of decay; usually bacteria and fungi.

dendrites: The short extensions of the neuron.

deoxyribonucleic acid: See *DNA*.

deoxyribose: The five-carbon carbohydrate attached to purine or pyrimidine bases within DNA molecules.

dermal tissue: The tissue that functions to protect the plant from injury and water loss and covers the outside of the plant.

diffusion: The movement of molecules from a region of high concentration to a region of low concentration, down its concentration gradient.

diploid: Cells having two sets of chromosomes.

diploid nuclei: Contained within a mass of cytoplasm within cellular slime molds.

disaccharides: Sugars composed of two molecules.

divisions: See *phyla*.

DNA: Deoxyribonucleic acid; a double-helix nucleotide molecule containing deoxyribose, nitrogenous base, and a phosphate group; contains the genetic information from which amino acids are determined.

DNA fingerprinting: A technique that uses electrophoresis to match DNA molecules to one another for identification purposes.

DNA polymerase: In DNA replication, the enzyme that adds complementary bases (nucleotides) to the exposed DNA template strand.

DNA replication: The process by which cells replicate or synthesize their DNA; takes place during S phase of the cell division cycle.

domain: The largest and broadest taxonomic category.

domestic breeding: A process of artificial selection that brings about new forms that differ from ancestral stock.

dominant allele: An allele that is expressed in a phenotype.

ductless glands: Glands that have no ducts, such as the endocrine glands.

duodenum: The first 10 to 12 inches of the small intestine in which most of the chemical digestion takes place.

eardrum: The tympanic membrane that receives vibrations from the outer ear and transmits them to the three inner ear bones.

ecosystems: Systems formed from the interactions between communities and their abiotic physical environments.

ectoderm: One of three germ layers that develops into the skin and nervous system.

egg cell: The haploid sex cell within the female ovary.

elements: The fundamental building blocks of matter within all living things.

embryo: The fetal structure that forms when all the organs of the body have taken shape.

embryology: The study of embryonic development.

endergonic reaction: Chemical reaction in which energy is obtained and trapped from the environment.

endocrine glands: Glands throughout the animal body that secrete hormones that help coordinate body systems.

endocytosis: The form of active transport in which a small patch of plasma membrane encloses particles that are near the cell surface.

endoderm: One of three germ layers that develops into the gastrointestinal tract.

endoplasmic reticulum (ER): An organelle composed of a series of membranes extending throughout the cytoplasm; two types exist: rough ER and smooth ER.

endoskeleton: An internal support system in the echinoderms and most vertebrates that may include a framework of bones and cartilage that serves as a point of attachment for muscle.

endosperm: The female tissue that encloses the seed within the angiosperms.

entropy: The degree of disorder or randomness of a system.

environmental fitness: An individual's ability to adapt to an environment and reproduce.

enzymes: Proteins that catalyze the chemical reactions within cells.

eosinophils: White blood cells whose functions are related to parasitic infections.

epididymis: The tube in which sperm cells mature.

epigenetic inheritance: Inheritable traits based on DNA modification, not the DNA sequence itself.

epiglottis: A thin flap of tissue that folds over the opening to the mammalian trachea during swallowing and prevents food from entering the trachea.

epinephrine: A hormone produced in the adrenal medulla that increases heart rate, blood pressure, and the blood supply to skeletal muscle.

erythrocytes: Red blood cells; disk-shaped cells produced in the bone marrow that have no nucleus; their cytoplasm is filled with hemoglobin to transport oxygen.

erythropoietin: A hormone produced by the kidney cells that functions in the production of red blood cells.

esophagus: A thick-walled muscular tube located behind the windpipe that extends through the neck and chest to the stomach.

estrogen: A hormone produced by the ovaries that stimulates the development of secondary female characteristics.

eubacteria: A kingdom of bacteria.

eukaryotes: Cells that contain a nucleus and internal cellular bodies called organelles.

evolution: Changes in inherited characteristics that occur within populations over time.

exergonic reaction: A chemical reaction in which energy is released.

exocrine glands: Glands, such as the salivary glands, that deliver their enzymes via ducts.

exocytosis: The reverse of endocytosis; internally produced substances are enclosed in vesicles and fuse with the cell membrane, releasing the contents to the exterior of the cell.

exoskeleton: The hard, protective, outer covering of arthropods and mollusks.

facilitated diffusion: The movement of molecules across a membrane from a region of high concentration to a region of low concentration that is assisted by proteins.

fallopian tubes: The passageways that egg cells enter after they are released from the ovaries; also called *oviducts.*

family: Similar genera classified together.

fats: Lipids composed of a glycerol and fatty acids.

fatty acids: Long chains of carbon atoms with carboxyl groups at one end.

feeding pattern: The pattern in which animals obtain their nutrients.

fermentation: An anaerobic process in which energy can be released from glucose even though oxygen is not available; occurs in yeast cells.

fertilized egg cell: An egg cell that has been fertilized by a sperm cell; called a *zygote.*

fetus: Results from a developing embryo at about 8 weeks when the embryo is somewhat human looking and the remaining development consists chiefly of growth and maturation.

flavin adenine dinucleotide (FAD): A coenzyme that functions in the production of ATP.

food chain: The transfer of food energy from producers to consumers.

food pyramid: A way of expressing the availability of food in an ecosystem at a successive number of trophic levels.

food web: Many interwoven food chains.

forebrain: A portion of the brain that consists of the cerebrum, thalamus, hypothalamus, and limbic system.

Fungi: A kingdom that includes the yeasts, molds, mildews, and mushrooms.

G_1 phase: A phase within interphase of the cell division cycle that prepares cells for DNA replication and growth.

G_2 phase: A phase within interphase of the cell division cycle that prepares cells for mitosis.

gametes: Sex cells of parent organisms; haploid cells.

ganglia: Knotlike groups of cell bodies that make up part of the autonomic nervous system.

gastrin: A hormone produced by digestive glands to influence digestive processes.

genes: The functional segments of chromosomes. Genes code for proteins.

gene flow: A mechanism of evolution that results when individuals migrate from one group to another and contribute their genes to the gene pool of the new population.

gene linkage: The concept of transfer of a linkage group.

gene linkage map: A map that pinpoints the location of genes based on their connection to certain marker gene sequences.

gene pool: The collection of genes within a population; as changes in the gene pool occur, a population evolves.

genetic code: Information coded within the DNA that stores the information for protein synthesis.

genetic drift: A mechanism of evolution that occurs when a small group of individuals leaves a population and establishes a new one in a geographically isolated region.

genome: The set of all genes that specify an organism's traits.

genomics: The study of an organism's entire set of chromosomes, including their function and species evolution.

genotype: The gene composition of a living organism.

genus: A grouping of similar species (plural, *genera*).

geographic distribution: The distribution of species in geographic areas.

geotropism: The turning of a plant away from or toward the earth.

gills: Structures that allow fishes to exchange gases with their environment.

glial cells: The cells of the nervous system that support, protect, and nourish the neurons.

glomerulus: A ball of capillaries that comprises the Bowman's capsule in the human kidney.

glottis: A slitlike structure at the opening to the mammalian trachea.

glucagon: A hormone produced in the pancreas that stimulates the breakdown of glycogen to glucose in the liver.

glucose: A carbohydrate with the chemical formula $C_6H_{12}O_6$ that serves as the primary carbon source of living things.

glycogen: A polysaccharide composed of thousands of glucose units that serves as the storage form of glucose in the human liver.

glycolysis: The subdivision of cellular respiration in which glucose molecules are broken down to form pyruvic acid molecules.

Golgi apparatus: An organelle within eukaryotic cells composed of a series of flattened sacs; the site of protein and lipid processing and packaging; also called *Golgi body*.

Graafian follicle: A cluster of cells within the ovary that is derived from egg cells and secretes the female hormone estrogen.

ground tissue: The tissue of the vascular plant that is responsible for storing the carbohydrates produced by the plant.

gymnosperms: Vascular plants having unprotected seeds, such as the conifers.

halophiles: Members of domain Archaea that can withstand highly saline conditions.

haploid: Cells containing one copy of each chromosome.

hemoglobin: A red pigmented protein that loosely binds to oxygen and carbon dioxide molecules and carries them through the bloodstream.

herbivores: Animals that eat plants.

heterotrophs: Species that acquire food from organic matter.

heterozygous: Two different alleles that are present for a particular characteristic.

hindbrain: The portion of the brain that consists of the medulla, pons, and cerebellum.

histone acetylation: Mechanism to facilitate gene expression. The enzymatic addition of acetyl groups to tightly bound DNA sequences will release the gene for transcription.

histones: Nuclear proteins that coil DNA molecules.

homeostasis: The process in which the internal environment exists at a steady-state equilibrium despite changes in the external environment.

homeothermic: The ability to maintain a constant body temperature, regardless of the animal's environmental temperature.

homozygous: Two identical alleles that are present for a particular characteristic.

hormones: Biochemical substances produced within plant or animal cells, or glands, that exert a particular effect.

hydrostatic skeleton: A water-based skeleton present in many animals (such as the earthworm) that lack structures, such as bone, for muscles to pull against.

hypothalamus: The portion of the forebrain that serves as the control center for hunger, thirst, body temperature, and blood pressure.

hypothesis: The proposal of a solution to the question within the scientific method.

ileum: The final 12 feet of the small intestine.

immune response: The stimulation of B- and T-lymphocytes.

incomplete dominance: An allele combination in which two characteristics appear to "blend" and both alleles can express themselves; one example is red, white, and pink snapdragons.

inner cell mass: A group of cells that continues to develop at one end of the blastocyst.

interneuron: A type of neuron that connects sensory and motor neurons and carries stimuli in the brain and spinal cord; also called *association neurons.*

interphase: The cell division cycle phase in which the cell spends most of its time; includes G_1, S phase (DNA replication), and G_2.

invertebrates: The most primitive of the chordates; lack a backbone.

involuntary muscle: See *smooth muscle* and *cardiac muscle.*

islets of Langerhans: Clusters of cells that make up the endocrine portion of the pancreas.

jejunum: The second 10 inches of the small intestine.

kinetochore: A region of DNA that has remained undivided during prophase of mitosis; binds to the spindle fibers that eventually pull apart the sister chromatids.

kingdom: The largest and broadest category of the classification system other than the domain.

Krebs cycle: The subdivision of cellular respiration in which pyruvic acid is broken down and the energy in its molecules is used to form high-energy compounds.

larynx: The "voice box" of mammals, formed from several folds of cartilage at the superior end of the trachea.

left atrium: The chamber of the human heart that receives oxygen-rich blood via the pulmonary vein.

left ventricle: The chamber of the human heart in which oxygen-rich blood enters through the bicuspid valve that leads into the aorta.

lens: The portion of the eye that focuses the light on the retina.

leukocytes: White blood cells produced in the bone marrow; have various functions in the body, such as immune reaction.

lichens: Formed by a combination of two organisms, fungi and cyanobacteria, living in a mutualistic relationship.

ligaments: The tough, fibrous tissues that link bones to one another.

limbic system: A collection of structures that ring the edge of the brain and function as centers of emotion.

lipid: An organic molecule used to form cellular and organelle membranes, the sheaths surrounding nerve fibers, and certain hormones; includes fats as an energy source.

liver: Organ that helps process the products of human digestion and removes excess glucose from the bloodstream, converting it to a polymer called glycogen for storage.

lungs: Pair of organs where oxygen diffuses into the blood to join with hemoglobin in the red blood cells.

lymph: A watery fluid derived from plasma that seeps out of the blood system capillaries and mingles with the cells.

lymph nodes: Capsulelike bodies that contain cells that filter the lymph and phagocytize foreign particles.

lymphatic system: The extension of the circulatory system consisting of capillaries called lymph vessels, a fluid called lymph, and structures called lymph nodes.

lymphatic vessels: A series of vessels that return the lymph fluid to the circulatory system.

lymphocytes: Certain white blood cells that are essential components of the immune system.

mammals: Milk-producing animals.

marsupials: Mammals whose embryos develop within the mother's uterus for a short period of time before birth. After birth, the babies complete their development in their mother's abdominal pouch.

medulla: The inner portion of the adrenal glands; a swelling at the tip of the hindbrain that serves as the passageway for nerves extending to and from the brain.

meiosis: The process by which the chromosome number is halved during gamete formation.

menstruation: The process by which the endometrium is released in females.

meristematic tissue: The growth tissue; the location of most cell division of vascular plants.

messenger RNA: See *mRNA*.

mesoderm: One of three germ layers that develops to become the muscles and other internal organs.

metabolism: The rapid turnover of chemical materials; involves the release or use of chemical energy.

metaphase: The stage during mitosis in which the replicated chromosomes align on the equatorial plate.

metaphase I: The phase during meiosis in which tetrads align on the equatorial plate.

metaphase II: The phase during meiosis II in which the chromatid pairs align at the center of the cell prior to separation.

methanogens: Members of domain Archaea that are strict anaerobes and create CH_4 (methane) as a waste product.

microbiome: The numerous bacterial communities found on and within the human body.

microRNA: See *miRNA*.

midbrain: A portion of the brain that lies between the hindbrain and the forebrain that consists of a collection of crossing nerve tracts.

minerals: Types of nutrients that include phosphorus, sulfur, potassium, magnesium, and zinc.

miRNA: MicroRNA; small, single-stranded RNA molecules that block translation of mRNA.

mitochondrion: The organelle that is the site of energy production in eukaryotic cells.

mitosis: The phase of the cell cycle during which the cell divides into two daughter cells.

molecule: Two or more atoms connected by chemical bonds.

Monera: The outdated kingdom that included all prokaryotic cells, now categorized as kingdoms Bacteria and Archaea.

monocytes: Some of the white blood cells that function in phagocytosis.

monosaccharides: Simple sugars that are composed of a single sugar molecule; one example is glucose.

monotremes: Egg-laying mammals that produce milk.

morula: A solid mass of cells that develops about 6 days after fertilization of an egg cell.

motor neuron: A type of neuron that transmits impulses from the brain and spinal cord to muscles or glands.

mRNA: Messenger RNA; the RNA molecules that receive the genetic code in the DNA and carry the code to ribosomes, where protein synthesis takes place.

multiple alleles: A condition in which more than two alleles exist for a characteristic; one example is A, B, AB, and O blood types.

muscle contraction: A process in which actin and myosin proteins move within a sarcomere.

mutation: A random change in the gene pool of a population that gives rise to new alleles and is the source of variation in a population.

mutualism: A symbiosis in which both partners benefit.

myelin sheath: A fatty layer of material that covers the axons of nerve cells.

myofibrils: Microscopic filaments that make up a muscle cell.

myosin: A protein microfilament that comprises the sarcomere of muscle cells.

natural selection: The concept that random, small variations take place in living things that lead to species evolution.

negative feedback: A control mechanism that strives to maintain homeostasis by reducing a stimulus.

nephron: The functional and structural unit of the kidney that produces urine and is the primary unit of homeostasis in the human body.

nerve cord: See *spinal cord.*

nerve impulse: An electrochemical event that occurs within the neuron.

nerve roots: The 31 pairs of projections that extend out along each side of the spinal cord; the sites of axons of the sensory and motor neurons.

nerves: Bundles of axons bound together.

neuroglia: The glial cells, together with the extracellular tissue.

neuron: A nerve cell.

neurotransmitter: A chemical substance that accumulates in the synapse and increases the membrane permeability of the next dendrite.

neutrophils: White blood cells that function as phagocytes, attacking and engulfing invading microorganisms.

nicotinamide adenine dinucleotide (NAD): A coenzyme that functions during respiration to produce ATP.

nicotinamide adenine dinucleotide phosphate (NADP): A coenzyme that functions during photosynthesis to produce ATP.

nitrogenous base: The nitrogen-based molecules that make up DNA (and RNA) molecules; two major types are purines and pyrimidines.

nonvascular plants: Plants that do not have specialized tissues to transport materials.

norepinephrine: A hormone produced in the adrenal medulla that intensifies the effects of epinephrine.

notochord: A flexible rod of tissue extending the length of an animal that provides internal support.

nucleic acids: Large molecules (DNA and RNA, for example) composed of nucleotides.

nucleoli: The small organelles that make up the nucleus; the site for ribosomal synthesis, assembly, and packaging (singular, *nucleolus*).

nucleotide: The unit that makes up nucleic acid; contains a nitrogen base, a phosphate group, and a carbohydrate molecule.

nucleus: The organelle within eukaryotic cells that contains the genetic material, DNA.

Okazaki fragments: New sections of DNA that are placed along the lagging strand during DNA replication and are joined together by DNA ligase to produce a new DNA strand.

olfactory nerve: The nerve that carries impulses from the nose to the brain for interpretation.

omnivores: Animals that consume both plants and animals.

oocytes: The developed oogonia in a human female after the age of puberty.

oogonia: Egg cells that accumulate in the ovaries before a human female is born.

optic nerve: The nerve that carries impulses from the eye to the brain for interpretation.

order: A grouping of similar families.

organelles: Microscopic bodies within a cell that perform distinct functions.

osmoregulation: Animal processes of regulating solute concentrations and balancing water gain and loss.

osmosis: The movement of water molecules across a membrane from a region of high concentration to a region of low concentration.

ovary: Female reproductive organ that contains the egg cells for reproduction; an endocrine gland that secretes estrogens. In plants, the structure of the pistil where the ovules are enclosed.

oviducts: See *fallopian tubes*.

ovulation: The process by which an egg cell is released from the follicle and swept into the fallopian tube, where it moves toward the uterus.

ovules: The protective structures that contain egg cells produced by the female.

oxytocin: A hormone released by the posterior pituitary gland that stimulates uterine contractions.

Pacinian corpuscles: The touch and pain receptors in the skin, muscles, and tendons.

paleontology: The science of locating, cataloging, and interpreting the life forms that existed in past millennia.

pancreas: A large, glandular organ lying near the stomach that produces many of the enzymes used to digest food.

parasites: Organisms that attack living things and cause disease.

parasitism: A type of symbiosis in which one organism benefits while the other is harmed.

parasympathetic nervous system: A subdivision of the autonomic nervous system that returns the body to normal after an emergency.

parathyroid glands: Glands located on the posterior surfaces of the thyroid gland that produce parathyroid hormone.

pathogenic: An organism that causes human disease.

PCR: Polymerase chain reaction; a technique used to amplify a gene of interest.

peptides: Small proteins.

peripheral nervous system: A collection of nerves that connect the brain and spinal cord to other parts of the body and the external environment.

peristalsis: A rhythmic series of muscular contractions that propels the bolus along.

peroxisome: Cytoplasmic body containing enzymes for digestion.

phagocytes: Cells that attack and engulf invading microorganisms.

phagocytosis: The process by which cells or microorganisms are engulfed by another cell; occurs when the vesicle formed from endocytosis contains solid matter.

pharynx: The cavity at the rear of the mouth that the nasal chambers open into; the throat.

phenotype: The expression of genes and the physical characteristics that result.

phloem: Structures of vascular plants that transport sugars and other nutrients from the leaves to the other parts of the plant.

phosphate group: A group derived from a molecule of phosphoric acid that connects the DNA molecules to one another.

phosphate ion: Combined with ADP, forms adenosine triphosphate (ATP).

photosystem: The site within the chloroplast in which sunlight is captured; includes the pigment molecules, proton pumps, enzymes, coenzymes, and cytochromes.

phototropism: The bending and turning of the plant stem toward a light source.

phyla: Related classes grouped together (singular, *phylum*).

physical map: A map that locates a gene of interest precisely by showing the actual number of base pairs between genes on a chromosome.

pineal gland: A human endocrine gland in the midbrain that regulates mating behaviors and day-night cycles.

pinocytosis: When the vesicle formed from endocytosis contains fluid.

pistil: The structure of the flower that contains a stigma, a style, and an ovary.

pith: The structure at the center of the stem of vascular plants.

pituitary gland: A gland at the base of the brain consisting of the anterior and posterior lobes that secretes several hormones.

placenta: The structure that supplies the fetus with nourishment.

placental mammals: Mammals that have a nutritive connection between the embryo and the mother's uterine wall.

Plantae: The kingdom that includes plants.

plant hormones: Hormones that regulate the growth and development of many plants.

plasma: A straw-colored liquid composed primarily of water; the fluid portion of blood.

plasma cells: Large antibody-producing cells derived from B-lymphocytes when stimulated.

plasma membrane: Also called a *cell membrane;* a membrane composed of lipids, proteins, and phospholipids.

plasmid: Small circular DNA molecules in bacteria often used as vectors to transform specific genes into cells.

plasmodial slime molds: Protists that consist of a single, flat, very large cell with many nuclei.

platelets: Small disk-shaped blood fragments produced in the bone marrow that serve as the starting material for blood clotting.

polygenic inheritance: The condition in which some characteristics are determined by an interaction of genes on several chromosomes or at several places on one chromosome; one example is human skin color.

polymerase chain reaction: See *PCR.*

polysaccharides: Complex carbohydrates formed by linking multiple monosaccharides.

pons: The portion of the hindbrain below the medulla and the midbrain that acts as a bridge between various portions of the brain.

population: An interbreeding group of individuals of one species occupying a defined geographic area.

positive feedback: A control mechanism that strives to amplify a stimulus, leading to a greater response.

predation: A relationship in which one population within a community may capture and feed upon another population.

producers: Autotrophic organisms within an ecosystem that trap energy (by photosynthesis).

progesterone: A hormone produced by the corpus luteum that regulates the buildup of tissue in the endometrium and inhibits the contractions of the uterus.

prokaryotes: Cells that do not contain a nucleus or internal organelles; include bacteria, cyanobacteria, and archaebacteria.

prophase: The first phase of mitosis; involves chromosomal condensation, nuclear membrane breakdown, and the migration of centrioles to opposite poles.

prophase I: The first phase of meiotic division, during which crossing over takes place.

prophase II: The phase during meiosis II in which the chromatin material condenses and each chromosome contains two sister chromatids attached by the centromere.

prostaglandins: The hormones secreted by various tissue cells that produce their effects on smooth muscles, on various glands, and in reproductive physiology.

proteinoids: The primitive polymers formed by the unison of amino acids; able to act as enzymes and catalyze organic reactions.

proteins: Long chains of amino acid units that are the main molecules from which living things are constructed.

Protista: A kingdom that includes protozoa, algae, and slime molds.

protocells: A perspective that posits that the first cells on Earth may have resulted from the clustering of macromolecules.

pulmonary artery: The artery of the human circulatory system that pumps the blood from the right ventricle to the lungs for gas exchange.

pulmonary vein: The vein of the human circulatory system that returns oxygen-rich blood from the lungs to the left atrium.

purine: A type of nitrogenous base present in DNA molecules containing two fused rings of carbon and nitrogen atoms; two examples in DNA are adenine (A) and guanine (G).

Purkinje fibers: The nerves that transfer amplified impulses to regions of the heart to control its function.

pyrimidine: A type of nitrogenous base in DNA molecules that has one ring containing carbon and nitrogen atoms; two examples in DNA are cytosine (C) and thymine (T).

recessive allele: The allele overshadowed by the dominant allele.

recombinant DNA: DNA molecules that have been altered in some way during the process of genetic engineering or biotechnology.

red blood cells: Also known as *erythrocytes*; cells that contain hemoglobin to transport oxygen.

reflex arc: The simplest unit of nervous activity; involved in the detection of a stimulus in the environment by sensory nerve endings, followed by impulses that travel via the sensory neurons to the spinal cord.

renal arteries: Arteries through which blood enters the kidneys.

renal veins: Veins through which blood exits the kidneys.

responsiveness: The ability of living things to respond to stimuli in the external environment.

resting potential: The inactive state of a neuron in which the cytoplasm is negatively charged with respect to the outside of the cell.

restriction enzymes: Enzymes that catalyze the opening of a DNA molecule at a "restriction" point; many leave dangling ends of DNA molecules at the point where the DNA has been opened.

retina: A single layer containing nerve cells within the eye that are light sensitive.

RFLP: Restriction fragment length polymorphism; a technique using small bits of DNA fragments linked to various diseases.

rhodopsin: A light-sensitive pigment of the eye that functions in the detection of light.

ribonucleic acid: See *RNA*.

ribosomal RNA: See *rRNA*.

ribosomes: Organelle bodies that may be bound to the endoplasmic reticulum (ER) that are the sites of protein synthesis in eukaryotic cells; the structures in which amino acids are bound together to form proteins.

right atrium: The chamber of the human heart in which oxygen-poor blood enters through a major vein called the vena cava.

right ventricle: One of the pumping chambers of the human heart from which blood exits; sends deoxygenated blood to the lungs.

RNA: Ribonucleic acid; a nucleic acid produced during transcription that is complementary to a DNA strand; similar to DNA in structure but contains the carbohydrate ribose and the pyrimidine uracil rather than thymine.

RNA polymerase: The enzyme that moves along the DNA strand, reads the nucleotides one by one, and synthesizes a complementary mRNA molecule according to the principle of complementary base pairing.

rod cells: The cells of the eye that permit vision in dim light.

roots: The structures of vascular plants that anchor them to the ground and take in water and minerals from the soil.

rough endoplasmic reticulum: ER studded with ribosomes; the site of protein synthesis in eukaryotic cells.

rRNA: Ribosomal RNA; RNA molecules that function to manufacture ribosomes.

salivary glands: The parotid glands, the submaxillary glands, and the sublingual glands that secrete saliva into the mouth.

sarcolemma: The muscle cell membrane.

sarcomere: The functional unit of the muscle that contains thin actin filaments and thick myosin filaments.

scientific method: An orderly process of gaining information about the biological world.

scrotum: A pouch outside the male body that contains the testes.

secretin: A hormone produced by digestive glands that influences digestive processes.

seedless vascular plants: The division Pteridophyta that includes the ferns.

semen: A fluid secretion containing sperm and secretions from the prostate gland, seminal vesicles, and Cowper's glands.

semilunar valves: Two valves found in the pulmonary artery and the aorta.

seminiferous tubules: Coiled passageways in which sperm production takes place.

sensory neurons: Neurons that receive stimuli from the external environment.

sensory somatic system: A subdivision of the peripheral nervous system that carries impulses from the external environment and the senses.

sepals: Modified leaves that enclose and protect a growing bud in flowers.

serum: Plasma from which the clotting proteins have been removed.

sex chromosomes: One pair among the 23 pairs of human chromosomes; the X and Y chromosomes.

siRNA: Small interfering RNA; small single-stranded RNA molecules that interfere with gene expression by binding complementary mRNA sequences.

skeletal muscle: See *striated muscle.*

slime molds (cellular): Amoebalike cells that live independently but may unite with other cellular slime molds to form an aggregate that functions as one unit.

small interfering RNA: See *siRNA.*

small intestine: A site of chemical digestion; includes the duodenum, jejunum, and ileum.

smooth endoplasmic reticulum: ER with no ribosomes attached.

smooth muscle: Found in the linings of the blood vessels, along the gastrointestinal tract, in the respiratory tract, and in the urinary bladder; contains few actin and myosin filaments; also called *involuntary muscle.*

species: A group of individuals that share features and are able to interbreed under natural conditions to yield fertile offspring.

spermatogonia: Cells within the seminiferous tubules that undergo a series of changes and then meiosis to yield sperm cells.

sperm cells: Haploid cells within the male testes.

S phase: The phase within the cell division cycle in which DNA is replicated.

spinal cord: The white cord of tissue passing through the bony tunnel made by the vertebrae.

spiracles: A series of openings on the body surface of terrestrial arthropods that open into tiny air tubes that assist in gas exchange.

spirilla: Flexible spiral bacteria (singular, *spirillum*).

spirochetes: Rigid spiral bacteria.

spleen: The organ where red blood cells are destroyed; a reserve blood supply for the body.

stamen: The structure of a flower that contains a thin, stemlike filament and an anther.

stem cell: A cell that has not yet differentiated into a specific cell type and retains the genetic ability to form any tissue.

stomata: The pores within leaves surrounded by guard cells that regulate the rate of gas exchange, which regulates the rate of photosynthesis (singular, *stoma*).

striated muscle: Skeletal muscle fiber that appears to be banded due to the presence of overlapping actin and myosin filaments; also called *voluntary muscle.*

substrate: The substance changed or acted on by an enzyme.

survival of the fittest: The concept of natural selection that states that the fittest survive and spread their traits through a population.

sutures: The immovable joints where bones come together within the skull.

symbiosis: The relationship between two populations that live together in a close, permanent, and mutually beneficial association.

sympathetic nervous system: A subdivision of the autonomic nervous system that prepares the body for an emergency.

synapse: The fluid-filled space separating the end of the axon from the dendrite of the next neuron or from a muscle cell.

synapsis: The process in which homologous chromosomes migrate toward one another and join to form a tetrad.

synergism: A type of relationship in which two populations accomplish together what neither could accomplish on its own.

telophase: A phase during mitosis in which the chromosomes arrive at the opposite poles of the cell.

telophase I: The phase during meiosis in which the nucleus reorganizes as the chromosomes become chromatin.

telophase II: The phase during meiosis II in which the chromosomes gather at the poles of the cells and form a mass of chromatin; the nuclear envelope develops, and the nucleoli reappear.

tendons: The connective tissue by which muscles are attached to bones.

testes: Endocrine glands that secrete androgens; the male reproductive organs located in the scrotum.

thalamus: A portion of the forebrain that integrates sensory impulses.

theory: A hypothesis that is confirmed through repeated experimentation and serves to explain phenomena.

thermophiles: Members of domain Archaea that thrive at temperatures that would otherwise kill cells; found in deep-sea volcanic vents and hot springs.

thrombocytes: The starting material for blood clotting; also called *platelets*.

thylakoids: Membranes that make up the grana in chloroplasts; the actual site of photosynthesis within chloroplasts.

thymosins: Hormones secreted by the thymus gland that influence the development of the T-lymphocytes of the immune system.

thymus gland: An endocrine gland in the neck tissues that secretes thymosins.

thyroid gland: A gland at the base of the neck that produces several hormones, such as thyroxine and calcitonin.

T-lymphocytes: White blood cells in the lymph nodes that are stimulated by microorganisms or other foreign material in the blood; also called T cells.

trachea: The windpipe of mammals.

tracheae: The branching network that extends from holes to all parts of an arthropod body to assist in gas exchange.

tracheids: The main conducting vessels of the xylem in most vascular plants.

tracheophytes: Vascular plants composed of xylem and phloem.

transcription: The process in which a complementary strand of mRNA is synthesized according to the nitrogenous base code of DNA.

transfer RNA: See *tRNA*.

transgenic: An organism that contains additional genes from another organism.

transgenic animals: Animals in which one or more genes have been introduced into the nonreproductive cells.

translation: The process by which the genetic code, in the form of mRNA, is translated to construct a protein.

tricuspid valve: A valve that passes blood from the right atrium into the right ventricle.

tRNA: Transfer RNA; RNA molecules in the cytoplasm of a cell that carry amino acids to the ribosomes for protein synthesis.

trophoblast: A layer of cells that forms after fertilization; projections from the trophoblast form vessels, which merge with maternal blood vessels to form the placenta.

tropism: The bending or turning response of a plant caused by external stimuli.

turgor pressure: The pressure exerted on a plant's guard cells to open.

umbilical cord: The source of attachment of the fetus to the maternal blood supply contained within the placenta.

urea: A component of urine that results from amino acid breakdown in the liver.

ureters: Tubes that carry waste from the kidneys to the urinary bladder for storage or release.

urethra: The path in which urine flows from the bladder to the exterior; the tube within the penis that carries the sperm.

uric acid: A component of urine that results from nucleic acid breakdown.

urinary bladder: The site where waste products are shipped from the kidney for storage or release.

urine: The product of the kidney; a watery solution of waste products, salts, organic compounds, uric acid, and urea.

uterus: A muscular organ in the pelvic cavity of female mammals; also called the *womb.*

vacuole: An organelle found in mature plant cells that stores nutrients and toxic waste.

vagina: A muscular organ in female mammals leading from the cervix to the exterior.

vascular bundles: Arrangements of the xylem and phloem in vascular plants.

vascular plants: Plants that contain specialized tissues to transport fluids.

vascular plants with protected seeds: Angiosperms; the most developed and complex vascular plants.

vascular plants with unprotected seeds: Gymnosperms; vascular plants that contain naked seeds, such as the conifers.

vascular tissue: Composed of phloem (sugar transport) and xylem (water and mineral transport).

vectors: The carriers of genes to be inserted into cells.

veins: Channels through which blood flows toward the heart.

vena cava: The major vein in the human heart; pumps oxygen-poor blood into the right atrium.

ventricle: A pumping chamber for blood to exit the heart.

vertebrates: Animals with backbones.

vessels: The main conducting vessels of the xylem found in the angiosperms.

virus: Fragments of nucleic acid surrounded by a protein coat; may attack cells and replicate within the cells, destroying them.

vitamins: Organic nutrients essential in trace amounts to the health of animals.

voluntary muscle: See *striated muscle.*

white blood cells: See *leukocytes.*

xylem: The structure of vascular plants that conducts water and minerals upward from the roots.

zygote: A fertilized egg cell, which is diploid.

INDEX

B

D

E

H

I

M

O

P

S

U

V

Made in the USA
Middletown, DE
08 January 2025

69085019R00156